The Making of Marx's 'Capital'

Roman Rosdolsky

THE MAKING OF MARX'S 'CAPITAL'

IN TWO VOLUMES

Translated by Pete Burgess

VOLUME 2

Pluto Press

First published in Germany by Europaische
Verlagsanstalt GmbH as *Entstehungsgeschichte des
Marxschen 'Kapital'*
Copyright © Europäische Verlagsanstalt,
Frankfurt am Main 1968
This translation first published 1977 by Pluto Press
11-21 Northdown Street, London N1 9BN
First paperback (abridged) edition published 1980
by Pluto Press
Second (unabridged) paperback edition published
1989 by Pluto Press in two volumes

Copyright © Pluto Press 1977

ISBN 0 86104 915 2 vol 1 paperback
ISBN 0 86104 305 7 vol 2 paperback

Contents

x · *Contents*

PART FOUR
The Section on the Circulation Process

Introductory Remark

We now come to a section of the *Rough Draft* which, in its most general sense, corresponds to Volume II of *Capital*, i.e. the section dealing with the circulation process of capital.

We should first of all note that when, in the chapter before last, reference was made to capital as it has 'become' (in contrast to capital 'in its becoming'), this was only an anticipation of results which were to follow much later in the analysis, as we have not yet by any means got beyond the stage of capital 'in its becoming'! This is because the 'completed form of capital' presupposes not only that capital has passed through the process of its own production, but also through that of circulation. In this sense, circulation represents a necessary moment in the self-formation (*Selbstgestaltung*) of capital – it is 'at the same time its becoming, its growth, its vital process'.[1] However, we can only speak of capital 'which has become' or is 'complete' when capital, 'steps, so to speak, beyond its organic inner life, and enters into relations with outer life',[2] that is, when the analysis progresses from that of 'capital in general' to that of 'many capitals', 'capital in its reality'.

It follows from this, that this section can only concern itself with

---

[1] *Grundrisse*, p.517.
[2] *Capital* III, p.44.

a general study of the circulation process : with the new forms which capital develops during its circuit, and in particular, during its stay in the sphere of circulation. We should not for one instant forget, however, that 'in reality this sphere represents the sphere of competition . . . which, considered in each individual case, is dominated by chance; where, then, the inner law, which prevails in these accidents and regulates them . . . remains, therefore, invisible and unintelligible to the individual agents in production'.[3] However, it is precisely for this reason that the scientific analysis of the circulation process must, in the first instance, disregard all the outward appearances of competition, so that it can grasp this process in its pure form, its 'simple basic form'.

Marx develops the concept of the circulation of capital from two standpoints. First, capital's stay in the sphere of circulation proper, i.e. in the markets for labour and commodities. Second, the circuit of capital through the entirety of its phases, which includes the phase of its production process, as well as the phase of circulation proper. Different characteristic forms emerge from each of these standpoints, which modify the laws arrived at in the previous section. Their analysis is therefore indispensable for the understanding of the process of capitalist production as a whole.

[3] *ibid.* p.828. Cf. also *ibid.* pp.43-44.

21.

The Transition from the Production Process of Capital to the Circulation Process of Capital. Excursus on the Realisation Problem and the First Scheme of Reproduction

In contrast to Volume II of *Capital*, the section of the *Rough Draft* which deals with the circulation process of capital, opens with an excursus, which, strictly interpreted, goes beyond the limits of the abstract analysis of the process of circulation and the new characteristic forms of capital which arise there. However, this section should be regarded as a welcome complement to the analysis. This excursus, which can be found on pages 401-23 deals with the problem of realisation and crises of overproduction.

It begins : 'We have now seen how, in the *valorisation process*, capital has (1) maintained its value by means of exchange itself . . . (2) increased, created a surplus-value. There now appears, as the result of this unity of the process of production and the process of valorisation, the product of the process, i.e. capital itself, emerging as product from the process whose presupposition it was . . . and specifically [as] a *higher value*, because it contains more objectified labour than the value which formed the point of departure. This value as such is *money*. However, this is the case only *in itself*; it is not *posited* as such;[1] that which is posited at the outset, which is on hand, is a commodity with a certain (ideal) price, i.e. which exists only ideally as a certain sum of money, and which first has to *realise* itself as such in the exchange process, hence has to re-enter the process of simple circulation in order to be posited as *money*.'[2]

Marx continues : 'Looked at precisely, the *valorisation process* of capital therefore appears at the same time as its *devaluation process*, its demonetisation.' This is because money has lost its form as money through its entry into the production process, and can only regain it in the circulation process. In fact, the situation is now that 'the capitalist enters the process of circulation not simply as one engaged in exchange, but as *producer*, and the others engaged in exchange are, in relation to him, *consumers*. They must exchange money in

---

[1] Cf. Note 71 on p.127 above.
[2] *Grundrisse*, pp.401-02.

order to obtain his commodity for their consumption, while he exchanges his product to obtain their money. Suppose that this process breaks down – and the separation by itself implies the possibility of such a miscarriage in the individual case – then the capitalist's money has been transformed into a worthless product, and has not only not gained a new value, but also lost its original value. But whether this is so or not, in any case devaluation forms one moment of the valorisation process;[3] which is already simply implied in the fact that the product of the process in its immediate form is not *value*, but first has to enter anew into circulation in order to be realised as such. Therefore while capital is reproduced as value and new value in the production process, it is at the same time posited as *not-value*, as something which first has to be *valorised by means of exchange* ... As a *commodity*, capital now shares the fate of commodities in general; it is a matter of accident whether or not it is exchanged for money, whether its *price* is realised or not.'[4] We thus come to the realisation problem, and by extension to the problem of crises. The *Rough Draft* states further : 'In the production process itself – where capital continued to be presupposed as value – its *valorisation* appeared totally dependent solely on the relation of itself as objectified labour to living labour; i.e. on the relation of capital to wage-labour. But now, as a product, as a commodity, it appears dependent on circulation, which lies outside this process ... As a commodity it must be (1) a use-value, and as such an object of need, object of consumption; and (2) it must be exchanged for its equivalent in money. The new value can be realised only through a sale.'[5]

Both of these conditions have been encountered in the analysis of simple commodity circulation. We saw there that : 'As a means of exchange the commodity must be a use-value, but it can only become such by being put up for sale – since it is not a use-value for the person in whose hands it is a commodity, but for the person who

---

[3] This devaluation (Marx uses the word in this sense only in the *Rough Draft*), which constitutes a moment of the process of valorisation itself, should be distinguished from the devaluation of capital which results from the increase in the productive power of labour. 'Value', we read in Marx's letter to Engels of 14 September 1851, 'is originally determined by the original costs of production. . . . But once it is produced, the price of the product is determined by the costs which are necessary to reproduce it. And the costs of reproduction fall constantly, and all the more rapidly, the more industrial is the age. Hence, law of the progressive devaluation of capital-value itself . . .' (*MEW* Vol.27, p.313.)

[4] *Grundrisse*, p.403.

[5] *ibid.* p.404.

takes it in exchange as a use-value. For the owner of the commodity its use-value consists solely in its exchangeability, alienability to the extent of the exchange-value represented in it.'[6] 'One and the same relation must therefore be simultaneously a relation of essentially equal commodities which differ only in magnitude, i.e. a relation which expresses their equality as materialisations of universal labour-time, and at the same it must be their relation as qualitatively different things, as distinct use-values for distinct needs, in short a relation which differentiates them as actual use-values.'[7] This contradiction between use-value and exchange-value, already manifest in the commodity and in simple commodity circulation, is revived in new forms in the circulation of capital. 'But this time, this contradiction is posited not merely as it was in (simple) circulation, as a merely formal difference;[8] rather the quality of being measured by use-value is here firmly determined as the quality of being measured by the sum total of the needs of the exchangers for this product . . . [so that] what is posited now is that the measure of its availability is given in its *natural composition itself*. In order to be transposed into the general form [i.e. the form of money] the use-value [produced by capital] has to be present in a limited and specific quantity; a *quantity* whose measure does not lie in the amount of *labour objectified in it*, but arises from its *nature as use-value*, in particular *use-value for others*.'[9] That is : 'As a *specific, one-sided qualitative* use-value, e.g. grain', the product of capital, 'is only required in a specific quantity; i.e. in a certain *measure*. This measure, however, is given partly in its quality as use-value – its *specific* usefulness, applicability – partly in the number of individuals engaged in exchange who have a need for this specific consumption. (The number of consumers multiplied by the magnitude of their need for this *specific* product.)' This is because 'use-value in itself does not have the boundlessness of value as such. Given objects can be consumed as objects of need only up to a certain level . . . Hence as *use-value*, the product contains a barrier – precisely the barrier consisting of the need for it – which, however, is measured not by the need of the producers but by the sum total of the needs of the exchangers.'[10] However, if this sum total of needs falls short, the

[6] *Grundrisse*, German edn. p.927.

[7] *Contribution*, p.44.

[8] In the sense that the commodity must undergo the change of form C-M and M-C.

[9] *Grundrisse*, p.406.

[10] Here Marx already touches on a theme we shall encounter once again in Volume III of *Capital* (pp.185, 194-95, 635-36).

product of capital ceases to be use-value, and consequently ceases to be capital.[11]

However, the product of capital must not only be an object of consumption, but must also 'be exchanged for its equivalent – in money',[12] which produces a further barrier to realisation. Since 'circulation was presupposed at the outset as a constant magnitude – as having a given volume – but since, on the other hand, capital has created a new value in the production process, it seems indeed as if no equivalent were available for it.' 'The surplus-value . . . requires a surplus equivalent', but this must first be created by production. Thus, capital has 'its limit as value . . . in alien production, just as use-value [has] its barrier in alien consumption; in the latter, its measure is the amount of need for the specific product, in the former, the amount of *objectified labour* existing in circulation.' 'The indifference of value as such towards use-value is thereby brought into just as false a position as are, on the other side, the substance of value and its measure as objectified labour in general.'[13]

Following on from this we find the illuminating description, which we have already encountered,[14] of the 'propagandising' and 'civilising' tendencies of capital, where Marx shows how capital's insatiable drive for valorisation brings about 'an extending circle of circulation . . . through production itself', and 'the continuous creation of more surplus labour . . . as a complement to itself'; and how, on the other side, this same drive for valorisation has as its consequence the fact that 'the consuming circle within circulation expands as did the productive circle previously', by means of the creation of new needs and the extension of existing ones. In this way it begins to look as if the impediments to realisation which have just been described could in fact be overcome through the development of the capitalist mode of production itself. However, it by no means follows that because capital tries to tear down every barrier which it encounters, 'and hence gets ideally beyond it', in terms of the direction in which it proceeds that 'it has *really* overcome it', that they cease to exist.[15] On the contrary : if we previously saw in capital 'a unity of the processes of production and valorisation,' it is now necessary to emphasise 'that this unity of production and valorisation is not *direct*, but is only a *process*' . . . by which all the contradictions

[11] *Grundrisse*, p.405.
[12] *ibid.* p.404.
[13] *ibid.* pp.405-07.
[14] Cf. Chapter 15 above.
[15] *Grundrisse*, p.410.

inherent in it 'are continually superseded ' ('overcome by force . . . although this supersession appears up to a certain point merely as a smooth restoration of the balance'), 'but also constantly recreated'.[16] How this process actually takes place is 'another question', which does not belong to the study of 'capital in general'. What is necessary at this stage of the analysis is 'to take note of the existence of these contradictions in the first instance', and demonstrate that both these contradictions and the tendencies which temporarily overcome them are in themselves already contained in the 'simple *concept* of capital' – so that their later unfolding can be seen as a development from this kernel.[17]

In the *Rough Draft* Marx only deals with the problem of crises of overproduction at this abstract level. He says : 'The whole dispute as to whether overproduction is possible and necessary in capitalist production revolves around the point whether the process of the valorisation of capital within production directly posits its valorisation in circulation, whether its valorisation posited in the *production process* is its *real* valorisation.' Bourgeois economists are divided into two camps on this issue. Those 'who like Ricardo, conceived production as directly identical with the self-valorisation of capital – and hence were heedless of the barriers to consumption or of the existing barriers of circulation itself, to the extent that it must represent counter-values at all points, having in view only the development of the forces of production and the growth of the industrial population – supply without regard to demand – have . . . grasped the positive essence of capital more correctly and deeply than those who, like Sismondi, emphasised the barriers of consumption and of the available circle of counter-values, although the latter has better grasped the limited nature of production based on capital in its negative one-sidedness.' Of course Ricardo also had 'a suspicion that the *exchange-value* of a commodity is not a value apart from exchange, and that it proves itself as a value only in exchange; but he regards the barriers which production thereby encounters as accidental, as barriers which are overcome. He therefore conceives the overcoming of such barriers as being in the essence of capital, although he often becomes absurd in the exposition of that view; while Sismondi, by contrast, emphasises not only the encounters with the barriers but their creation by capital itself, and has a vague intuition that they must lead to its breakdown. He therefore wants to put up barriers to production from the out-side, through custom, law etc. which of course, as merely external

[16] *ibid.* pp.406, 407.
[17] *ibid.* pp.36off.

and artificial barriers, would necessarily be demolished by capital. On the other side, Ricardo and his entire school never understood the really *modern crises*, in which this contradiction of capital discharges itself in great thunderstorms which increasingly threaten it as the foundation of society and of production itself.'[18]

It is already clear from this comparison of Sismondi's and Ricardo's views which direction Marx's solution to the problem had to take. In opposition to Ricardo he constantly points to the 'fundamental contradiction' of capitalism; to 'the poverty and restricted consumption of the masses, as opposed to the drive of capitalist production to develop the productive forces as though only the absolute consuming power of society constituted their limit'.[19] Thus, for Marx, crises are not, 'as Ricardo maintains, accidental . . . but essential outbreaks – occurring on a large scale and at definite periods – of the immanent contradictions'.[20] However, in opposition to Sismondi, Marx stresses the 'universal tendency' and 'positive essence' of capital, and consequently views the (periodic) overcoming of the 'barriers of the sphere of exchange' as part of the 'essence of capital' itself in the sense which we are already acquainted with, according to which the contradictions of capitalism are continually 'superseded', but equally continually 'posited' on an ever-higher scale, until they eventually lead to its collapse and the transition to a 'higher form of social production'.

This is precisely the standpoint from which Marx criticises the attempts at apologetics by the economists (James Mill, Say, Ricardo, MacCulloch) who deny the possibility of general crises of overproduction. Since we are already acquainted with this critique from Volume II of the *Theories*, it is unnecessary to look in any more detail at the relevant passage in the *Grundrisse* (411-414). Marx's main criticism of the economists was that they disregarded all the features, and definitions specific to the capitalist mode of production in order to 'explain away' crises of overproduction, and that they equated the circulation of capital with simple commodity circulation, even direct barter. 'The moment of valorisation is here simply thrown out entirely, and production and consumption are simply equated, i.e. not production based on capital but production based directly on *use-value* is presupposed.'[21] On the other hand, bourgeois

---

[18] *ibid*. pp.410-11. Cf. the well-known account of Sismondi's views in *Theories* III, pp.55-56. There too, Sismondi is contrasted with Ricardo.
[19] *Capital* III, p.484.
[20] *Theories* III, p.56.
[21] *Grundrisse*, p.413.

economists always strive to emphasise only the moment of unity, and to deny the antagonisms themselves, whereas in fact, 'the economic relation . . . comprises antagonism . . . and is the unity of opposites'. In this way 'the unity of antagonisms becomes the immediate identity of these antagonisms'[22] – 'a manner of thinking the criticism of which belongs to the sphere of logic and not of economics'.[23] Thus, the economists stressed, for example, 'the tendency of capital to distribute itself in correct proportions (between the different branches of production)', but studiously forgot, that 'it is equally its necessary tendency – since it strives limitlessly for surplus labour, surplus productivity, surplus consumption etc. – to drive beyond this proportion.'[24] Of course, if production in capitalism were carried out in accordance with a general and predetermined plan, 'then no over-production could in fact occur'.[25] However, since this is a contradiction in terms, as the growth of capitalist production 'is not directly regulated or determined by the needs of society', capital is necessarily 'just as much the constant positing as the supersession of proportionate production',[26] and within capitalist production proportionality proceeds 'as a continuous process out of disproportionality'.[27] Bourgeois apologetics should therefore be countered by saying that, although the individual moments of the valorisation process belong internally together, 'they may or may not find each other, balance each other, correspond to each other' – and that 'their indifferent,

---

[22] *Theories* III, pp.88, 101. Cf. *Theories* II, pp.500-01 : 'If, for example, purchase and sale – or the metamorphosis of commodities – represent the unity of two processes, or rather the movement of one process through two opposite phases, and thus essentially the unity of two phases, the movement is essentially just as much the separation of these two phases and their becoming independent of each other. Since, however, they belong together, the independence of the two correlated aspects can only show itself forcibly as a destructive process. It is just the *crisis* in which they assert their unity, the unity of the different aspects. The independence which these two linked and complementary phases assume in relation to each other is forcibly destroyed. Thus the crisis manifests the unity of the two phases that have become independent of each other. There would be no crisis without this inner unity of factors that are apparently indifferent to each other. But no, says the apologetic economist. Because there is this unity, there can be no crises. Which in turn means nothing but that the unity of contradictory factors excludes contradiction.'
[23] *Contribution*, p.96.
[24] *Grundrisse*, p.413.
[25] *Theories* III, p.118.
[26] *Grundrisse*, p.414.
[27] *Capital* III, p.257.

independent existence towards one another is already a foundation of contradictions' (and crises).[28]

'Still', continues Marx, 'we are by no means finished. The contradiction between production and valorisation – of which capital, by its concept, is the unity – has to be grasped more intrinsically than merely as the indifferent, seemingly reciprocally independent appearance of the individual moments of the process, or rather of the totality of processes.' That is, this has not been achieved by mere reference to the general, abstract possibility of crises; rather it is necessary 'to demonstrate that capital contains a *particular* restriction of production – which contradicts its general tendency to drive beyond every barrier to production'. This already suffices, 'in order to have uncovered the foundation of *overproduction*, the fundamental contradiction of developed capital, or, to speak more generally, to have uncovered the fact that capital is not, as the economists believe, the *absolute* form for the development of the forces of production.'[29] This particular limitation consists in the fact that capital's drive for valorisation, which compels it to expand production without limit (i.e. without regard to the available market or effective demand), at the same time forces it to restrict the sphere of exchange, 'i.e. the possibility of valorising, of realising the value posited in the production process'. This contradiction is one which Sismondi grasped 'crudely, but nonetheless correctly . . . as a contradiction between production for the sake of production and distribution which makes absolute development of productivity impossible'.[30]

Marx now proceeds to substantiate his theory in more detail. He says that it is a basic presupposition of capitalist production that capital has to enter into exchange with the worker, before anything else, i.e. posit necessary *labour*. 'Only in this way does it valorise itself and create surplus-value'. On the other hand, 'it posits necessary labour only *to the extent* and *insofar* as it is *surplus labour* and the latter is *realisable as surplus-value*. It posits surplus labour, then, as the condition for necessary, and surplus-value as the limit of objectified labour, of value as such . . . It therefore restricts labour and the creation of value . . . and does so on the same grounds as and to the same extent that it posits surplus labour and surplus-value. By its nature therefore, it posits a *barrier* to labour and value-creation, in contradiction to its tendency to expand them boundlessly. And in as much as it both posits a barrier *specific* to itself, and on the other side

[28] *Grundrisse*, pp.414-15.
[29] *ibid.* p.415.
[30] *Theories* III, p.84.

equally drives over and beyond *every* barrier, it is the living contra-diction.'[31]

We read in the next paragraph that if capital 'makes surplus labour and its *exchange* for *surplus labour* into the precondition of necessary labour . . . hence already narrows and attaches conditions to the sphere of exchange from this side – it is just as essential to it, on the other side, to restrict the worker's consumption to the amount necessary to reproduce his labour-capacity – to make the value which expresses *necessary labour* the barrier to the realisation of labour-capacity and hence of the worker's *exchange capacity*, and to strive to reduce the relation of this necessary labour to surplus labour to the minimum'.[32] This is a tendency which arises from capital's limit-less drive for valorisation, but which must, in fact, once more, appear as a restriction in its sphere of exchange.

The same situation prevails with the productive force. 'On the one hand, the necessary tendency of capital to raise it, to the utmost, in order to increase *relative surplus labour-time*. On the other hand, thereby decreases necessary labour-time, hence the worker's exchange capacity. Further . . . relative surplus-value rises much more slowly than the force of production, and moreover this proportion grows ever smaller as the magnitude reached by the productive forces is greater.[33] *But the mass of products grows in a similar proportion*' as the productive force does. 'But to the same degree as the mass of products grows, so grows the difficulty of realising the labour-time contained in them – because the demands made on consumption rise.'[34]

Marx sums up by saying : 'Capital, then, posits *necessary labour-time* as the barrier to the exchange-value of living labour-capacity; *surplus labour-time* as the barrier to necessary labour-time; and *surplus-value* as the barrier to surplus labour-time; while at the same time it drives over and beyond all these barriers', forgets them and abstracts from them. 'Hence overproduction i.e. the sudden *recall*[35] of all these necessary moments of production founded on capital; hence general devaluation in consequence of forgetting them. Capital, at the same time [is] thereby faced with the task of launching its attempt anew from a higher level of the development of productive

---

[31] *Grundrisse*, p.421.
[32] *ibid.* pp.421-22.
[33] Cf. Chapter 16 above.
[34] *Grundrisse*, p.422.
[35] This is reminiscent of Hegelian terminology. (Cf. G.Lukacs, *The Young Hegel*, pp.542-43.)

forces, with each time greater collapse *as capital*. Clear therefore, that the higher the development of capital, the more it appears as barrier to production – hence also to consumption – besides the other contradictions which make it appear as burdensome barrier to production and intercourse.'[36]

The contradiction between production and valorisation therefore lies in the nature of capital itself and is based upon the antithetical interaction between necessary and surplus labour. The larger the surplus labour, the smaller (relatively) the necessary labour; but also then, the smaller the possibility of the realisation of the surplus-product. In this sense capital's limitless drive for valorisation is 'identical to the positing of barriers to the sphere of exchange'.[37]

But if this is the case, if capital itself erects a barrier to the realisation of the surplus-value created in the production process by limiting workers' consumption, how is the development of capital possible at all? How can capital avoid a situation of permanent crisis? If one proceeds from these premises, isn't it then necessary to accept the explanation offered by Sismondi or the Russian Narodniks, namely, that in the long run the realisation of surplus-value is impossible in capitalism – and that surplus-value has to be disposed of abroad, in exchange with other countries?[38]

In the first instance we want to confine ourselves to a methodological criticism of this argument. That is, that those who argue in this way overlook the fact that in reality the capitalist economy presupposes competition – that is a sphere in which the abstract characteristics acquired in the analysis of 'capital in general' are first realised, but are at the same time modified ('mediated').

This connection is clear from Marx's *Rough Draft*. We read in the excursus that in contrast to the conditions prevailing before capitalism, in capitalist production 'consumption is mediated at all points by exchange, and labour never has a *direct* use-value for those who are working. Its entire basis is labour as exchange-value and as the creation of exchange-value.' Consequently the wage-labourer (as distinct from the producers of previous periods) is 'himself an independent centre of circulation, someone who exchanges, posits exchange-value, and maintains exchange-value through exchange'. But precisely for this reason, 'to each capitalist, the total mass of all workers, with the exception of his own workers, appear not as workers, but as consumers, possessors of exchange-values (wages), money,

[36] *Grundrisse*, pp.422-23, 416.
[37] *ibid.* p.422.
[38] See Chapter 30 below.

which they exchange for his commodity.[39] They form a proportionally very great part – although not quite so great as is generally
imagined, if one focuses on the industrial worker proper – of all consumers. The greater their number – the number of the industrial
population – and the mass of money at their disposal, the greater
the sphere of exchange for capital.' (And we know 'that it is the tendency of capital to increase the industrial population as much as
possible'.) However, 'the relation of one capitalist to the workers of
another capitalist . . . alters nothing in the relation of capital in
general to labour. Every capitalist knows this about his worker, that
he does not relate to him as producer to consumer, and (he therefore) wishes to restrict his consumption, i.e. his ability to exchange
his wage, as much as possible.[40] Of course he would like the workers
of *other* capitalists to be the greatest consumers possible of *his own
commodity*. But the relation of *every* capitalist to *his own workers* is
the *relation as such of capital and labour*, the essential relation.'
From this point of view, therefore, it is basically an 'illusion . . . –
true for the individual capitalist as distinct from all the others – that
apart from his workers the whole remaining working class confronts
him as *consumer* and *participant in exchange*, as money-spender,
and not as worker . . . It is forgotten that, as Malthus says,[41] "the very
existence of a profit upon any commodity presupposes *a demand
exterior to that of the labourer who has produced it*", and that therefore "the *demand of the labourer himself can never be an adequate*
demand". However, this illusion is of the greatest significance.

[39] 'What precisely distinguishes capital from the master-servant relation
is that the worker confronts him as a consumer and possessor of exchange-
values, and that in the form of the *possessor of money* in the form of money,
he becomes a simple centre of circulation – one of its infinitely many centres,
in which his specificity as worker is extinguished.' (*Grundrisse*, pp.420-21.)

[40] In opposition to this one could cite the example of the American
motor industry, which, as is known, is dominated by three giant firms who
certainly would want to regard the workers they employ as purchasers of their
cars. However, if there are several firms in one branch of industry, each of
them can and will hope to force their commodities onto the workers of their
competitors and will, therefore, aim to reduce the wages of their own workers
(and consequently their ability to exchange) 'as much as possible'.

[41] In fact it is not Malthus himself who says this, but rather Otter, the
editor of his *Principles*, in a footnote which runs: 'The demand created by
the productive labourer can never be an adequate demand, because it does
not go to the full extent of what he produces. If it did, there would be no
profit, consequently no motive to employ him. The very existence of a profit
upon any commodity presupposes a demand exterior to that of the labour
which has produced it. Ed.' (T.R.Malthus. *Principles of Political Economy
etc.*, 1836, p.405. See Grundrisse, p.418, editorial footnote.)

This is because, as we read further in the *Rough Draft*, 'one production sets the other into motion and hence creates consumers for itself in the *alien* capital's workers, it *seems* to each individual capitalist that the demand of the working class posited by production itself is an adequate demand. On one side, this demand which production itself posits drives it forwards, and must drive it forwards beyond the *proportion* in which it would have to produce with regard to the workers; on the other side, if the demand *exterior to the demand of the labourer himself* disappears or shrinks up then the collapse occurs. Capital itself then regards *demand by the worker* – i.e. the payment of the wages on which this demand rests – not as a gain, but as a loss, i.e. *the immanent relation between capital and labour* asserts itself. *Here again it is the competition among capitals*, their indifference to, and independence of, one another, which brings it about that the individual capital relates to the workers of the entire remaining capital *not as to workers* : hence is driven beyond the right proportion.'[42]

'It is quite the same', we read in a footnote, 'with the demand created by production itself for raw materials, semi-finished goods, machinery, means of communication, and for the auxiliary materials consumed in production, such as dyes, coal, grease, soap etc. This effective exchange-value-positing demand is adequate and sufficient as long as the producers exchange among themselves. Its inadequacy shows itself as soon as the final product encounters its limit in final and direct consumption.[43] This *semblance* too, which drives beyond the correct proportion, is founded in the essence of capital, which, as will be developed more closely in connection with competition, *is* something which repels itself, *is* many capitals mutually quite indifferent to one another.[44] Insofar as one capitalist *buys* from others, buys commodities, or sells, they are within the simple exchange relation; and do not relate to one another as capital. The *correct* (imaginary) proportion in which they must exchange with one another in order to realise themselves at the end as capital lies *outside* their relation to one another.'[45]

So much then on the 'necessary', 'immanent' barriers to capitalist production, as they are expressed in crises of overproduction. It is clear that we are not dealing here with absolute barriers, but with barriers which only evidence themselves as such in con-

[42] *Grundrisse*, p.420.
[43] Cf. *ibid*. pp.149, 639.
[44] Cf. Note 117 on p.42 above.
[45] *Grundrisse*, p.421.

tinuous movement, in the continual struggle of conflicting tendencies. If this is so then the question as to the conditions which facilitate a relative equilibrium of the capitalist system as it reproduces itself, though this may be interrupted by crises, is not only theoretically admissible, but in fact of very great importance for the science of economics.

Marx[46] divides the aggregate capital of society into five classes, which are represented by capitalists A, B, C, D, and E. The first two are manufacturers of raw materials, the third produces machines, the fourth necessaries for the workers, and the fifth luxury products destined for consumption by the capitalists themselves. The organic composition of capital is the same in all five branches of production : 75c + 25v (where c is divided into $\frac{2}{3}$ raw material, and $\frac{1}{3}$ machinery). The rate of exploitation is also the same – 100%. Thus we get the following scheme :

	Machinery	Raw Material	Labour	Surplus-Product
A. 1. Raw material manufacture	20	40	20	20
B. 2. Raw material manufacturer	20	40	20	20
C. Machinery manufacturer	20	40	20	20
E. Workers' necessaries[47]	20	40	20	20
D. Surplus-product	20	40	20	20

How does reproduction take place according to this scheme?* Capitalist E 'exchanges his entire product of 100 for 20 in his own workers' wages, 20 in wages for workers of raw material A, 20 for the workers of raw material B, 20 for the workers of machinery maker C, 20 for the workers of surplus producer D; of this he exchanges 40 for raw material, 20 for machinery, 20 he obtains back from workers' necessaries, and 20 remain for him to buy surplus produce, from which he himself lives. Similarly the others in the relation.' (i.e., each of the manufacturers of raw materials keeps 40 in raw material – since he can use this directly, without exchange, for new production[48] – and

---

[46] *ibid.* p.439.

[47] The scheme abstracts from fixed capital.

[48] It is assumed that this is a question of raw materials which can be used in the manufacture of raw materials again. (The same applies to Capitalist C's machines.)

* Translator's note : order follows *Grundrisse*, p.441.

exchanges 60 for the products of other capitalists; whereas the 'machinist' and the 'surplus producer' can only retain 20 each in machinery and luxury goods respectively – and each have 80 to exchange.) Each capitalist is then in a position to continue production at the same level in the next year.[49]

It is not difficult to recognise in this fivefold scheme in the *Rough Draft* the prototype of the schemes for simple reproduction which we know from *Capital*[50] and the *Theories*.[51] If we group together the two producers of raw materials and the 'machinist' on the one side, and on the other side the producers of necessaries and luxuries, thus forming two particular groups, we arrive at the following scheme :

	Machinery	Raw Material	Labour	Surplus-Product
I. Means of Industries Production	60	120	60	60
II. Consumption good industries	40	80	40	40

or :

$$\text{I.} \quad 180c + 60v + 60s$$
$$\text{II.} \quad 120c + 40v + 40s$$

The capitalists in Department I can directly use 180c for reproduction – since they exist in the natural form of means of production; and the 40v and 40s of Department II have just as little need to go outside the limits of this department. However, what has to be exchanged between the two departments are the 60v and 60s of the first, and the 120c of the second department. Thus we obtain the equation which expresses the conditions for the smooth course of reproduction as : $vI + sI = cII$ as in the reproduction scheme of Volume II of *Capital*.

However, what happens in the case of extended reproduction, or, expressing this in capitalist terms, in accumulation? (For if the five capitalists in the scheme 'consumed the entire surplus, they would be no further at the end than at the beginning, and the surplus-value

---

[49] *ibid.* pp.439-41.
[50] *Capital* II, pp.401-02; and *Capital* III, p.838.
[51] *Theories* III, pp.246-49. (Cf. *MEW* Vol.30, pp.362-67, and Marx's 'Tableau Economique' printed in the one-volume German edn. of *Capital*, 1948, pp.533-36.)

of their capital would not grow' – which would contradict the aim of capitalist production.) This requires a special scheme for extended reproduction and we can already find the hastily drafted beginning to such a scheme in the *Rough Draft*.[52] And although this attempt contains some obvious errors of calculation, it is clear what Marx wants to say. It proceeds as follows :

In the first place, the production of at least the 'surplus producer' D has to be restricted, so that a transition from simple to extended reproduction can take place. For, if each capitalist consumed only 10 out of 20 from his surplus-value and accumulated 10, then the surplus producer D can now only produce 50 (5 × 10) units of luxury commodities. (In this way the transition to extended reproduction is already bound up with crisis.) Secondly, however, each capitalist (A,B,C, and E) must employ 5 of the 10 accumulated units of value for raw material, $2\frac{1}{2}$ for machinery and $2\frac{1}{2}$ for wages, in the same proportions as before. Only if these proportions are kept to is there a *'real possibility* for greater valorisation – the production of new and larger values.' If not then the two capitalists E and D, the producers of necessaries and luxuries, would be producing too much – 'that is, too much relative to the proportion of the part of capital going to the worker, or too much relative to the part of capital consumable by the capitalists (too much relative to the proportion by which they must increase their capital . . .'). That is, a *'general over-production'* would occur, 'not because relatively *too little* [sic] had been produced of the commodities consumed by the workers or too little [sic] of those consumed by the capitalists, but because too much *of both* had been produced – *not* too much *for consumption,* but too much to retain the *correct proportion between consumption and valorisation*; too much for valorisation'.[53]

'In other words', continues Marx, 'at a given point in the development of the productive forces – for this will determine the relation of necessary labour to surplus labour – a fixed relation becomes established in which the product is divided into 4 parts – corresponding to raw material, machinery, necessary labour, surplus labour – and finally surplus labour into one part which goes to consumption and another which becomes capital again. This inner division, inherent in the concept of capital, appears in exchange in such a way that the exchange of capitals among one another takes place in specific and restricted proportions – even if these are constantly changing, in the course of production . . . This gives, in any

[52] *Grundrisse*, p.442.
[53] *ibid.* pp.442-43.

case, both the sum total of the exchange which can take place, and the proportions in which each of these capitals must both exchange and produce. If the relation of necessary labour to the constant part of capital is as, e.g. in the above example [i.e. as in the scheme] . . . then we have seen that the capital which works for the consumption of capitalists and workers combined may not be greater than $\frac{1}{5} + \frac{1}{10}$ of the 5 capitals . . . Given likewise is the relation in which each capital must exchange with each other one, which represents a specific one of its own moment. Finally in which each of them must exchange at all.'[54]

However, the point of the scheme is simply to show the 'inner', 'conceptual' division of capital – that is the conditions which make possible a situation of equilibrium in a capitalist system engaged in the process of growth. In reality these conditions for equilibrium can only be achieved in the face of continual disturbances. For, 'exchange in and for itself gives these conceptually opposite moments an indifferent being', so that they 'develop independently of one another'; consequently 'their inner necessity' only 'becomes manifest in the crisis, which puts a forcible end to their seeming indifference to one another'.[55]

However, this is not the only danger which threatens the valorisation of capital. Since the 'proportions for the exchange between capitals' are determined by the 'proportion between necessary labour and surplus labour', and since this proportion itself is dependent on the development of the productive forces, every 'revolution in the forces of production' must bring about a change in these proportions. If, despite this, production proceeds indifferently onwards (and capital's boundless drive for valorisation tends to drive it beyond all the 'correct' proportions!), 'then ultimately a minus, a negative magnitude, will come out of the exchange on one side or the other'. For 'the barrier always remains that exchange – hence production as well – takes place in such a way that the relation of surplus labour to necessary labour remains the same – for this is = to the constancy of the valorisation of capital.'[56] However, if production is driven beyond these limits, then at a definite moment in time a situation of 'general devaluation and destruction of capital' comes about. Thus the crisis resolves itself in 'an actual reduction of production, of living labour – in order to re-establish the correct proportion between

[54] *ibid.* p.443.
[55] *ibid.* pp.443-44. (The concept of indifference, which we repeatedly encounter in the *Rough Draft*, is also taken from Hegel's *Logic*.)
[56] *ibid.* p.444.

necessary and surplus labour, on which everything depends in the last instance.' 'Both are therefore posited in the essence of capital; the devaluation of capital . . . as well as the supersession of devaluation and the creation of the conditions for the valorisation of capital.'[57] This section has taught us two things : firstly, that Marx's much discussed schemes of reproduction only aim to show how – within limited periods of time, with relatively constant techniques of production and rate of exploitation of labour – extended reproduction can take place, given that definite proportions are maintained between the main departments of social production : and therefore that any 'harmonious' interpretation of this scheme is inapplicable. And secondly we can see from this excursus, above all, the stress which Marx placed on the contradiction between capital's boundless drive for valorisation, and the limited power of consumption of capitalist society. This is a point which many writers in the marxist tradition ignore, or only look at incidentally, although it is indispensable for the understanding of Marx's theory of crises. This is a theme we shall deal with in more detail in Part VII of this work.

[57] *ibid.* pp.446-47. Marx adds : 'The process by which this takes place in reality can be examined only as soon as real capital i.e. competitions etc. – the actual real conditions – have been examined.'

22.
Circulation Time and its Influence on the Determination of Value

The section of the *Rough Draft* which we described in the previous chapter basically represents simply the notification and preliminary treatment of a series of questions which were not to have been finally solved until a much later stage of the analysis, in fact not until after the completion of the *Rough Draft* itself.[1] Its aim was to indicate the barriers and difficulties in the realisation process, which followed from the study of 'capital in general', but which only existed as 'possibilities' and which consequently could only be 'overcome as possibilities'.

However, the main point of the section of the *Rough Draft* dealing with the circulation process is to 'represent the sphere of circulation . . . in relation to the characteristic forms which it produces', in order to demonstrate 'the development in the nature of capital, which takes place there'.[2] For this purpose, we have to assume, (as in the previous chapter), 'that capital passes through its process of circulation in the normal way', hence that – regardless of the extent of the difficulties of realisation – 'the capitalist must have succeeded in selling his commodities, and in reconverting the money shaken loose from them into capital'. This is by no means an arbitrary assumption, but corresponds 'to what actually takes place', insofar as the reproduction of capital does actually occur.[3]

The analysis of the production process has yielded the result, that the valorisation of capital consists exclusively in the appropriation of unpaid alien labour, and the extent of this valorisation is most precisely measured by the amount of surplus labour-time extracted

---

[1] We should remember that Marx's original plan relegated the treatment of crises to the last (sixth) book of the work.

[2] Marx saw this himself (*Capital* III, p.828) as the function and content of Volume II of *Capital*. Cf. *Grundrisse*, p.524 : 'Circulation as we regard it here is a process of transformation, a qualitative process of value . . . insofar as new aspects are created within this process of transformation as such – in the transition from one form to another.'

[3] *Capital* I, pp.709, 710 (564, 565).

from the workers. However, is this the only significance of the time factor in production? Shouldn't we also regard the entire time that capital remains in the process of production as value-creating and surplus-value-creating, even if this does not directly represent the period of labour?

We refer here to the distinction between the length of the production process itself – production-time – and the duration of the labour-time which is necessary for the manufacture of the product.[4] In agriculture for example (and to a greater or lesser extent in several other branches of production), 'there are interruptions given by the conditions of the production process itself, pauses in labour-time, which must be begun anew at the given point in order to continue or to complete the process; the constancy of the production process here does not coincide with the continuity of the labour process.' Or, 'after the product is finished it may be necessary for it to lie idle for some time, during which it needs relatively little labour, in order to be left in the care of natural processes, e.g. wine.'[5] Thus, it may be the case that the same labour-time may be expended for different products, and that, in fact, the production-time can exhibit noticeable differences, which – since they take the form of different turnover periods[6] for different entrepreneurs – have to be 'compensated for', if equal amounts of capital are to yield equal amounts of profit. But says Marx, 'this question obviously belongs only with equalisation of the rate of profit'.[7] However, he already wants to refute the idea that 'a natural circumstance which hinders[8] a capital in a specific branch of production from exchanging with the same amount of labour-time in the same amount of time as another capital in another branch of production can in any way contribute to *increasing* the former's value. Value, hence also surplus-value, is not = to the time which the production phase lasts, but rather to the labour-time, objectified and

---

[4] See *Capital* II, Chapter XIII. This distinction is only fleetingly dealt with in the *Rough Draft* – just sufficiently to show its influence on the valorisation of capital. One can also see, in the relevant pages of the manuscript, how Marx first elaborated this distinction. (For example, on p.518 production-time is still equated with labour-time, which was subsequently corrected by the insertion of the word 'false'.)

[5] *Grundrisse*, pp.602-3.

[6] See Chapter 23 below.

[7] *Grundrisse*, p.669.

[8] 'The non-identity of production-time with labour-time can be due generally only to natural conditions, which stand directly in the path of the valorisation of labour, i.e. the appropriation of surplus labour by capital. These obstacles in its path do not of course constitute advantages, but rather, from its point of view, losses.' (*Grundrisse*, p.670.)

M

living, employed during this production phase. The living labour-time alone – and, indeed, in the proportion in which it is employed relative to objectified labour-time – can create surplus-value, because [it creates] surplus labour-time . . .'[9] And it is for precisely this reason that it is impossible to attribute any value-creating role to production-time – as distinct from labour-time.

So much on the significance of the time factor, insofar as it applies to capital's stay in the sphere of production. However, it is still necessary for capital to spend time in the sphere of circulation, after the production phase is completed, which also takes time. What happens during this time : how does it affect the creation of value and the valorisation of capital?

In the first instance we should remain conscious of the fact that *'circulation proceeds in space and time'*. In this sense we have to distinguish between 'spatial' or 'real' circulation, and 'economic' circulation proper. The first – the physical bringing of products onto the market – 'belongs, economically considered . . . to the production process itself', and can be regarded 'as the transformation of the product into a *commodity*', since 'the product is really only finished when it is on the market. The movement through which it gets there belongs still with the cost of making it.'[10] In fact, transportation 'only changes the location of the product'. 'Whether I extract metals from mines, or take commodities to the site of their consumption, both movements are equally spatial.'[11] Transporting the product to market 'gives it a . . . new use-value (and this holds right down to and including the retail grocer, who weighs, measures, wraps the product and thus gives it a form for consumption),[12] and this new use-value costs labour-time, is therefore at the same time exchange-value.'[13] How-

9 *ibid.* p.669.

10 *ibid.* pp.533-34.

11 *ibid.* p.523. 'If one imagines the same capital both producing and transporting, then both acts fall within direct production, and circulation . . . would begin only when the product had been brought to its point of destination.' (*ibid.*)

12 Marx propounds the same standpoint in *Capital* III, Chapter XVII and *Capital* II, Chapter VI, Section II.

13 *Grundrisse*, p.635. However, when the transported commodity 'has reached its destination, this change which has taken place in its use-value has vanished, and is now only expressed in its higher exchange-value, in the enhanced price of the commodity. And although in this case the real labour has left no trace behind it in the use-value, it is nevertheless realised in the exchange-value of this material product; and so it is true also of this industry as of other spheres of material production that the labour incorporates itself in the commodity . . .' (*Theories* I, p.413.)

ever, from this standpoint transport does not constitute a 'special case', in contrast to direct production – although it is true that the transportation industry distinguishes itself from the other areas of investment of productive capital by the fact that 'it appears as a continuation of a process of production *within* the process of circulation and *for* the process of circulation'.[14]

In contrast to 'real' circulation, which brings the products to the site of their consumption and thereby makes them into commodities for the first time, actual 'economic' circulation is simply a 'qualitative process of value', 'the change of form which value undergoes as it passes through different phases'.[15] This circulation also requires time – namely 'the time it necessarily costs to transform the commodity into money and the money back into commodity'.[16] However, does there not enter, precisely through this, 'a moment of value-determination . . . independently of labour, not arising directly from it, but originating in circulation itself?'[17]

Marx's answer is that this certainly is the case. 'In as much as the renewal of production depends on the sale of the finished products', or 'the transformation of the commodity into money and re-transformation of money into the conditions of production', and in as much as the stay in the sphere of circulation constitutes a necessary part of the life of capital, then 'how many products can be produced in a given period of time; how often capital can be valorised in a given period of time, how often it can *reproduce* and *multiply* its value . . . (depends naturally), on the velocity of circulation, the time in which it is accomplished'. This 'is evidently a condition not posited directly by the production process itself'.[18] Thus, it is obvious at first glance that if, for example, a capital of, say, 100 thalers passes through 4 turnovers in a year, and each time yields a profit of 5 per cent, then this is the same (disregarding any possible accumulation), 'as if a capital 4 times as large, at the same percentage . . . were to turn over *once* in one year – each time 20 thalers'. (In the original : 20 per cent.) 'The velocity of turnover, therefore – the remaining conditions of production being held constant – substitutes for the *volume* of capital.'[19] In this sense 'the more frequent turnover of capital in a given period of time' is the same as 'the more frequent harvests during the natural year in the southerly countries compared

[14] *Capital* II, p.155.
[15] *Grundrisse*, pp.524, 626.
[16] *ibid.* p.625.
[17] *ibid.* p.519.
[18] *ibid.* pp.537-38.
[19] *ibid.* p.519.

with the northerly'.[20] Hence the velocity of circulation is of the greatest importance for capital since it is evident that the speed of the production process depends on it, and, as a consequence, 'if not values themselves', then 'the volume of values to a certain degree'.[21]

However, in what sense does circulation time affect the determination of value? Let us return to the example of the harvests. We spoke of countries where a favourable climate allows frequent harvests. However, if for example 'the real conditions of wheat production in a given country permit only one harvest, then no velocity of circulation can make two harvests out of it'. But if, in contrast, 'an obstruction in the circulation occurred, if the farmer could not sell his wheat soon enough . . . then production would be delayed', and with this the net profit of the one harvest would be endangered.[22] That is, the most which can be achieved by means of the acceleration of circulation is the avoidance of the impediments to reproduction which are inherent in the nature of capital itself. So, the circulation time of capital is nothing other than the time of its devaluation,[23] if the first is shortened, then the second is shortened too. What certainly cannot be concluded from this is that the valorisation of capital has thereby become larger; merely that its devaluation (*Entwertung*) has become smaller.

Marx says further : 'The difference shows itself simply in this : if the totality of labour-time commanded by capital is set at its maximum, say infinity ∞, so that necessary labour-time forms an infinitely small part and surplus labour-time an infinitely large part of this ∞, then this would be the maximum valorisation of capital, and this is the tendency towards which it strives. On the other side, if the *circulation time of capital* were $= 0$, if the various stages of its transformation proceeded as rapidly in reality as in the mind, then that would likewise be the maximum of the factor by which the production process could be repeated, i.e. the number of capital valorisation processes in a given period of time. The repetition of the production process would be restricted only by the amount of time it lasts, the amount of time which elapses during the transformation of raw material into product.' By contrast, 'if either surplus labour-

[20] *ibid.* p.519.
[21] *ibid.* p.538.
[22] *ibid.* p.544.
[23] Cf. the beginning of the previous chapter. 'Just as grain when it is put in the soil as seed loses its immediate use-value, is devalued as immediate use-value, so is capital devalued from the completion of the production process until its re-transformation into money and from there into capital again.' (*ibid.* p.519.)

time or necessary labour-time $=$ o i.e. if necessary labour-time absorbed all time, or if production could proceed altogether *without* labour then neither value, nor capital, nor value-creation would exist.'[24] 'It is clear, therefore, that circulation time, regarded absolutely, is a deduction from the maximum of valorisation, is $<$ absolute valorisation. It is therefore impossible for any velocity of circulation or any abbreviation of circulation to create a valorisation $>$ that posited by the production phase itself. The maximum that the velocity of circulation could effect, if it rose to ∞, would be to posit circulation time $=$ o, i.e. to abolish itself. It can therefore not be a positive value-creating moment, since its abolition – circulation without circulation time – would be the maximum of valorisation; its negation $=$ to the highest position of the productivity of capital.'[25] Rather, circulation time can only influence value-creation and capital valorisation in a negative way, in that its acceleration or deceleration serve to shorten or extend merely the time during which capital is unable to employ productive labour and valorise itself.[26] 'In this respect, circulation time adds nothing to value; . . . does not appear as value-positing time, the same as labour-time.'[27]

But what about the costs of circulation, the expenditure of living or objectified labour, which result from 'passing through the various economic moments as such'? In this case the general law applies, i.e. *'that all costs of circulation which arise only from changes in the forms of commodities do not add to their value.* They are merely expenses incurred in the realisation of the value or in its conversion from one form into another. The capital spent to meet those costs (including the labour done under its control) belongs among the *faux frais* of capitalist production. They must be replaced from the surplus-product and constitute, as far as the entire capitalist class is concerned, a deduction from the surplus-value or surplus-product, just as the time a worker needs for the purchase of his means of subsistence is lost time'.[28]

The *Rough Draft* illustrates this with the following example : 'If, of two individuals, each one were the producer of his own product, but their labour rested on division of labour, so that they exchanged with each other, and the valorisation of their product depended . . . on this exchange, then obviously the time which this

---

[24] *ibid.* pp.538-39.
[25] *ibid.* pp.629-30.
[26] Cf. *Capital* II, p.128.
[27] *Grundrisse*, p.626.
[28] *Capital* II, p.152.

exchange would cost them, e.g. the mutual bargaining, calculating before closing the deal, would not make the slightest addition either to their products or to the latter's exchange-values.[29] If A were to argue that the exchange takes up such and such a quantity of time, then B would respond in kind. Each of them loses just as much time in the exchange as the other. The exchange time is their common time. If A demanded 10 thalers for the product – its equivalent – and 10 thalers for the time it costs him to get the 10 thalers from B, then the latter would declare him a candidate for the madhouse.' This is because the loss of time, which both suffer through the acts of exchange arises simply 'from the division of labour, and the necessity of exchange',[30] and must, therefore, appear as a deduction from their productive activity. ('If A produced everything himself, then he would lose no part of his time in exchange with B, or in transforming his product into money and the money into product again.') However, if the producers were to find that 'they could save time by inserting a third person, C, as a middleman between them, who consumed his time in this *circulation process*', (of course this would be if not only A and B, but a larger number of producers were to do the same), then, 'each of them would have to cede . . . a share of his product to C. What they would gain thereby would only be *a greater or lesser loss.*'[31]

Marx concludes that for this reason the actual costs of circulation 'can never multiply value', 'are not reducible to productive labour-time'. They are the *faux frais* of commodity production, and as such are inseparable from the capitalist mode of production.[32] 'Merchant's trade and still more the money trade proper' are to be understood in this sense. Insofar as they reduce the costs of exchange by their intervention 'they add to production, not by *creating* value, but *by reducing the negation of created values* . . . If they enable the producers to create more values than they could without this division of labour, and, more precisely, so much more that a surplus remains

---

[29] 'If the owners of the commodities are not capitalists but independent direct producers, the time employed in buying and selling is a diminution of their labour-time, and for this reason such transactions used to be deferred (in ancient and medieval times) to holidays.' (*ibid.* p.133.)

[30] Marx later summarises his argument by saying: 'It is wrong, therefore, for J.St.Mill to regard the cost of circulation as necessary price of the division of labour. It is the cost only of the spontaneously arisen division of labour, which rests not on community of property, but on private property.' (*Grundrisse*, p.633.)

[31] *ibid.* pp.624-25, 633.

[32] *ibid.* pp.625, 633.

after the payment of this function, then they have in fact increased production. Values are then increased however, not because the operations of circulation have created value, but because they have absorbed less value than they would have done otherwise. But they are a necessary condition for capital's production.'[33]

But what about the time which the capitalist himself loses in exchange? Shouldn't it also be regarded as 'labour-time', and consequently as 'value-creating'? Not at all, since he is only a capitalist 'i.e. representative of capital, personified capital . . . by virtue of the fact that he relates to labour as alien labour, and appropriates and posits alien labour for himself . . . The fact that the worker must work surplus labour-time is identical with the fact that the capitalist does not need to work, and his time is thus posited as not-labour-time; that he does not work the *necessary* time either. The worker must work surplus time in order to be allowed to objectify . . . the labour-time necessary for his reproduction. On the other side, therefore, the *capitalist's necessary labour-time* is *free* time, not time required for direct subsistence.' And Marx says that it is precisely for this reason that the time which the capitalist employs for the exchange of the commodities produced by him, 'looked at economically', 'concerns us here exactly as much as the time he spends with his mistress'.[34] 'If time is money, then from the standpoint of capital it is only alien labour-time, which is of course in the most literal sense the capitalist's money.' Circulation time 'interrupts the time during which capital can appropriate alien labour-time, and it is clear that this relative devaluation of capital cannot add to its valorisation, but can only detract from it; or, insofar as circulation costs capital objectified alien labour-time, values. (For example because it has to pay someone who takes over this function.) In both cases, circulation time is of interest only insofar as it is the suspension, the negation of alien labour-time';[35] and in both cases it proves to be a barrier to the productivity of capital, and a deduction from surplus labour-time and from surplus-value.

However, aren't the differences in valorisation which are the product of differences in the circulation times of different capitals, equalised by the general rate of profit – in the same way as the difference between production-time and labour-time mentioned at the

[33] *ibid.* p.633.
[34] Marx states further on in the text, 'Otherwise, it could still be imagined that the capitalist draws compensation for the time during which he does not earn money as another capitalist's wage-labourer . . .'
[35] *Grundrisse*, p.634.

beginning of this chapter?[36] Certainly. 'As long as capital remains frozen in the form of the finished product, it cannot be active as capital, it is *negated* capital ... This thus appears as a loss for capital, as a relative loss of its value, for its value consists precisely in the valorisation process . . . Now let us imagine *many* capitals in particular branches of business, all of which are *necessary* (which would become evident if, in the eventuality of a massive flight of capital from a given branch, supply falling below demand, the market price would therefore rise above the natural price in that branch; i.e. above the price of production), and let a single branch of business require e.g. that capital A remain longer in the form of devaluation, i.e. that the time in which it passes through the various phases of circulation is longer than in all other branches of business – in which case this capital A would regard the smaller new value which it could produce as a positive loss, just as if it had so many more outlays to make in order to produce the same value. It would thus charge relatively more exchange-value for its products than the other capitals, in order to share the same rate of gain. But this could take place in fact only if the loss were distributed among the other capitals.'

Marx continues : 'Nothing more absurd, then, than to conclude that, because one capital obtains a compensation for its *exceptional* circulation time . . . now that all capitals [are] combined, *capital* can make something out of nothing, make a plus out of a minus – make a plus surplus-value out of a minus surplus-value . . . The manner in which the capitals among other things compute their proportional share of the *surplus-value* – not only according to the surplus labour-time which they set in motion, but also *in accordance with the time which their capital has worked as such* i.e. lain fallow, found itself in the phase of devaluation – does not of course alter in the least the total sum of the surplus-value which they have to distribute among themselves. This sum itself cannot grow by being smaller than it would have been if capital A, instead of lying fallow, had created surplus-value . . . And this *lying-fallow* is made good for capital A only insofar as it arises necessarily out of the conditions of the particular branch of production, and hence appears in respect to *capital as such*

---

[36] Marx deals with the question of the general rate of profit (or average rate) in several parts of the *Rough Draft*, although, as we already know, this theme did not, according to the original plan, come under the scope of 'capital in general', but under 'many capitals'. It is therefore no accident that in the final work the average rate of profit is first dealt with in Volume III, where the representation approximates more to the concrete forms of capital i.e. the sphere of competition. (See Chapter 25 below.)

as a burden on valorisation, as a *necessary barrier* to its valorisation generally.'[37]

And we read in another section of the *Rough Draft* : 'If one thinks of one capital, or one thinks of the various capitals of a country as one capital (national capital) as distinct from that of other countries,[38] then it is clear that the time during which this capital does not act as a productive capital i.e. posits no surplus-value, is a deduction from the valorisation time available to this capital. It appears as the negation not of the really posited valorisation time, but of the *possible* valorisation time, i.e. possible, if circulation time = o. It is clear, now, that the national capital cannot regard the time during which it does not multiply itself as time in which it does multiply itself, no more than e.g. an isolated peasant can regard the time during which he can neither harvest nor sow, during which his labour generally is interrupted, as time which makes him rich.' Marx adds : 'The fact that capital regards itself, and necessarily so, as productive and fruit-bearing independently of labour, of the absorption of labour', that it 'assumes itself as fertile at all times, and calculates its circulation time as value-creating time – as production cost – is quite another thing'.[39] However, the reason why this semblance arises, and must arise, will not be shown until we have studied the 'secondary process of valorisation', i.e. profit and the general rate of profit.[40]

One remark in conclusion. What has been stated in this chapter can also naturally be applied to money, and the circulation of money. We read in the *Rough Draft* : '*Money* itself, to the extent that it consists of precious metals, or its production generally – e.g. in paper circulation – creates expense, to the extent that it also costs labour-time, adds no value to the exchanged objects – to the exchange-values; rather its costs are a deduction from these values, a deduction which must be borne in proportional parts by the exchangers.'[41] And in another passage : 'Regarded in both of the aspects in which it occurs in the circulation of capital, both as medium of circulation and as the realised value of capital, money belongs among the costs of circulation insofar as it is itself labour-time employed to abbreviate circulation time on the one hand, and, on the other, to represent a

[37] *Grundrisse*, pp.546-48.
[38] Cf. pp.44-48 above.
[39] *Grundrisse*, p.662. Cf. *Capital* II, p.128 : 'But Political Economy sees only what is apparent, namely the effect of circulation time on capital's valorisation process in general. It takes this negative effect for a positive one, because its consequences are positive.'
[40] See Chapter 25 below.
[41] *Grundrisse*, p.625.

qualitative moment of circulation – the retransformation of capital into itself as value-for-itself. In neither aspect does it increase the value. In one aspect it is a precious form of representing value, i.e. a costly form, costing labour-time, hence representing a deduction from surplus-value. In the other aspect it can be regarded as a machine which saves circulation time, and hence frees time for production. But insofar as it itself, as such a machine, costs labour and is a product of labour, it represents for capital *faux frais de production*. It figures among the costs of circulation.' Hence capital's striving 'to suspend it in its inherited, immediate reality, and transform it into something merely *posited* and at the same time suspended by capital, into something purely ideal.'[42] We have already seen from Marx's remarks cited in Chapter 9 precisely why this tendency cannot be fully realised, and we shall return to this subject once more in the chapter on interest and profit.[43]

---

[42] *ibid.* pp.670-71. We read further in the text that, 'Supersession of money in its immediate form appears as a demand made by money circulation once it has become a moment of the circulation of capital; because in its immediate, presupposed form it is a barrier to the circulation of capital. The tendency of capital is circulation without circulation time; hence also the positing of the instruments which merely serve to abbreviate circulation time as mere formal aspects posited by it . . .' (*ibid.*)

[43] See Chapter 27 below.

23.

The Turnover of Capital and Turnover Time. The Continuity of Capitalist Production and the Division of Capital into Portions

We have already pointed out on several occasions that the life-span of capital is by no means confined to the actual process of production, but equally includes its circulation process. 'These form the two great sections of its movement, which appears as the totality of these two processes. On one side labour-time, on the other, circulation time. And the whole of the movement appears as unity of labour-time and circulation time, as unity of production and circulation. This unity itself is motion, process. Capital appears as this unity-in-process of production and circulation, a unity which can be regarded both as the totality of the process of its production, as well as the specific completion . . . of *one* movement returning into itself.'[1]

In other words, the circuit of capital – understood as the movement of capital through its various phases (from the advance of the capital-value to its return) – can be looked at in two ways : either as an individual, self-contained process, or as the same circuit in its periodicity, in its continual repetition. Marx adopted both methods of study in Volume II of the final work. The first was used in Part I of Volume II of *Capital*, where he was concerned with 'the forms which capital continually assumes and discards in its circuit' as well as 'the different forms of this circuit itself'.[2] (The fact that the circuit of capital was constantly repeated could contribute nothing substantial to the analysis at this point.) It was a different matter in the section which followed, Part II of Volume II, where Marx wanted to show how every industrial capital appears in the forms of productive capital, money-capital and commodity-capital, 'within the flow and succession of forms', 'simultaneously, if in varying degrees', where

[1] *Grundrisse*, p.620.

[2] *Capital* II, p.357. It should be mentioned here that the theme dealt with in Part I of Volume II ('The Metamorphoses of Capital and Their Circuits') – the reading of which presents such difficulties, but which surely represents a high point in the application of the dialectical method – is totally absent in the *Rough Draft*, which is a considerable weakness in the presentation of the circulation process there.

these forms, 'not only alternate with one another, but different portions of the total capital-value are constantly side by side and function in these different states'.[3] And this could only be represented if the circuit of capital was regarded not as one isolated segment, but as the totality of the movement of the capital-value-in-process.

We read in Volume II that 'a circuit performed by a capital and meant to be a periodical process, not an individual act, is called its turnover. The duration of this turnover is determined by the sum of its time of production and its time of circulation. This time total constitutes the time of turnover of the capital. It measures the interval of time between one circuit period of the entire capital-value and the next, the periodicity in the process of life of capital, or, if you like, the time of the renewal, the repetition, of the process of valorisation, or production, of one and the same capital-value.'[4]

What, then, is the significance of the turnover of capital in the circulation process of the capitalist economy?

The importance of this question will become particularly evident when we come to the representation of the specific types of turnover of fixed and circulating capital,[5] and a more precise definition of the average rate of profit.[6] It is sufficient here to recapitulate briefly what we established in the previous chapter.

Since the turnover time of capital is equal to the sum of its production-time and its circulation time, it is clear that differences in the duration of the turnover can originate from both factors – that is, from both the production-time and the circulation time.

As far as production-time is concerned, two facts are relevant here. In the first place, there are differences in the duration of labour which different products require for their production. One product may be completed within a week, another, perhaps, not until after several months – even if the labour-time which is employed daily in both cases is the same. This difference in the periods of labour[7] required for the production of the two products must, of course, also

---

[3] *ibid.* p.357.
[4] *ibid.* p.158.
[5] See Chapter 24 below.
[6] See Chapter 25 below.
[7] 'When we speak of a working day we mean the length of working time during which the worker must daily spend his labour-power, must work day by day. But when we speak of a working period we mean the number of connected working days required in a certain branch of industry for the manufacture of a finished product. In this case the product of every working day is but a partial one, which is further worked upon from day to day, and only at the end of the longer or shorter working period receives its finished form, is a finished use-value.' (*Capital* II, p.234.)

imply a difference in the turnover periods of the capitals concerned.[8] Secondly, we should refer to the difference between production-time and labour-time, which we have already met. The question here is of those interruptions to the production process which are 'independent of the length of the labour process, and conditioned by the nature of the product and its manufacture themselves'. During these interruptions 'the object of labour is subject to natural processes, which may take a shorter or longer time, in which it has to go through physical, chemical and physiological changes, during which the labour process is either totally or partially suspended'.[9] In this situation the production-time is greater than the labour-time, and it is clear that the turnover period of capital will be extended 'in accordance with the length of that production-time which does not consist of labour-time'.[10] And finally, we have the division into fixed and circulating capital which arises from the variation in the material forms in which productive capital exists, which results in the turnover of capital being subject to considerable modifications – as we shall see in the next chapter.

Even more important still are the variations in the periods of turnover which originate during the circulation phase. As we saw, 'the more rapid the circulation, the shorter the circulation time, the more often can the same capital repeat the production process. Hence, in a specific cycle of turnovers of capital, the sum of values created by it (hence surplus-value as well) . . . is *directly proportional to the labour-time and inversely proportional to the circulation time* . . . the total value = labour-time, multiplied by the number of turnovers of the capital.' Or, the value created by capital no longer seems to be simply determined by the labour employed in the production process, 'but rather by the coefficient of the production process; i.e. the number which expresses how often it is repeated in a given period of time'.[11] However, what follows from this is that even with capitals of the same magnitude, organic composition, and rate of surplus-value, the duration of the turnover period can be very different – hence that in this sense (as it states in the *Rough Draft*) the circulation time 'is itself a moment of production, or rather appears as a limit to production'.[12] However, the real concern of this chapter is something different, namely, a new contradiction of the capitalist

[8] See *Capital* II, Chapter XII.
[9] *ibid.* p.242.
[10] *ibid.* p.243.
[11] *Grundrisse*, p.627.
[12] *ibid.* p.628.

mode of production, which is revealed by the necessity of circulation and circulation time.

We saw that capital 'by its nature only preserves its character as capital in that it constantly functions as capital in the repeated process of production'.[13] Consequently, 'the *constant continuity* of the process, the unobstructed and fluid transition from one form into the other, or from one phase of the process into the next, appears as a fundamental condition for production based on capital to a much greater degree than for all earlier forms of production.'[14] Naturally, this continuity of production would best be ensured if there were no necessity at all for circulation time. However, this is impossible, since it is inherent in the nature of capital that it actually 'travels through the different phases of circulation not as it does in the mind, where one concept turns into the next at the speed of thought, in no time, but rather as situations which are separate in time. It must spend some time as a cocoon before it can take off as a butterfly. Thus the conditions of production arising out of the nature of capital itself contradict each other.'[15] They can only be mediated in practice (disregarding credit), 'by capital's dividing itself into parts, of which one *circulates as finished* product, and the other *reproduces itself in the production process*. These parts alternate; when one part returns into phase *P* (production process), the other departs. This process takes place daily, as well as at longer intervals . . . The whole capital and the total value are reproduced as soon as both parts have passed through the production process and circulation process, or as soon as the second part enters anew into circulation. The point of departure is thereby the terminal point. The turnover therefore depends on the size of the capital or rather . . . on the *total sum* of these two parts. Only when the total sum is reproduced has the entire *turnover* been completed; otherwise only $\frac{1}{2}$, $\frac{1}{3}$, $1/x$, depending on the proportion of the constantly circulating part.'[16]

'The question', continues Marx, 'is what part of the capital can now be continuously occupied in production (during the whole year)?' 'This matter must be reducible to a very simple equation, to which we shall return later[17] . . . This much is clear however. Call production time *pt*, circulation time *ct*. Capital *C*. *C* cannot be in its

[13] *ibid.* p.403.
[14] *ibid.* p.535.
[15] *ibid.* pp.548-49.
[16] *ibid.* p.661.
[17] See *Capital* II, Chapter XV (Effect of the Time of Turnover on the Magnitude of Capital Advanced).

production phase and its circulation phase at the same time. If it is to continue to produce while it circulates, then it must break into two parts, of which one in the production phase, while the other in the circulation phase, and the continuity of the process is maintained by part *a* being posited in the former aspect, part *b* in the latter. Let the portion which is always in production be *x*; then *x* = *C-b* (let *b* be the part of the capital always in circulation) . . . If *ct*, circulation time, were = o, then *b* likewise would be = o, and *x* = *C*. *b* (the part of the capital in circulation): *C* (the total capital) = *ct* (circulation time): *pt* (production-time); *b* : *C* = ct : *pt*; i.e. the relation of circulation time to production-time is the relation of the part of capital in circulation to the total capital.'[18]

Nevertheless, all that is achieved by the division of the capital into portions is that the whole capital does not have to interrupt its production process for the period of circulation – the continuity of the process is maintained. (If this were not the case, if the whole of the capital-value had to function first as money-capital, then as productive capital, and finally as commodity capital, then, instead of production being carried out continuously, it would take place 'in jerks and would be renewed only in periods of accidental duration according to whether the two stages of the process of circulation[19] were accomplished quickly or slowly,'[20] a state of affairs which is already ruled out by the technical basis of capitalist production). Despite this, the division of capital into portions cannot prevent some parts of capital from lying fallow in every capitalist undertaking, thus preventing its valorisation.[21] Hence capital's necessary tendency to cut circulation time by improving communications, developing the credit system etc. i.e. to establish 'a circulation without circulation

---

[18] *Grundrisse*, p.666.

[19] What are meant are the stages: M–C, or, more precisely, $M \underset{MP}{\overset{LP}{\diagdown}}$

(Purchase of labour-power and the means of production) and C'–M' (Retransformation of the capital-value expanded in production into its original money-form).

[20] *Capital* II, p.105.

[21] 'The effect of the turnover on the production of surplus-value and consequently of profit . . . briefly summarised . . . [is that] owing to the time span required for turnover, not all the capital can be employed all at once in production; some of the capital always lies idle, either in the form of money-capital, of raw material supplies, of finished but still unsold commodity-capital, or of outstanding claims; that the capital in active production, i.e. in the production and appropriation of surplus-value, is always short by this amount, and that the produced and appropriated surplus-value is always curtailed to the same extent.' (*Capital* III, p.70. This chapter was in fact written by Engels.)

time'. This is an aspect we shall return to in Chapter 27 'Fragments on Interest and Credit'.]

Because the turnover time of capital includes both labour-time and circulation time, nothing is easier than to credit to the latter what is in fact contributed by the former, thus attributing to capital 'a mystical spring of self-valorisation independent of its process of production, and hence of the exploitation of labour . . . which flows to it from the sphere of circulation'.[22] This conception forms the basis for most of the illusions of both the capitalists themselves and the bourgeois economists, who are ensnared in the capitalist manner of thinking.

[22] *Capital* II, p.128. (Cf. *Grundrisse*, p.640.)

24.
The Characteristic Forms of Fixed and Circulating (Fluid) Capital

I.

In his Preface to Volume III of *Capital* Engels refers to the common misunderstanding, according to which Marx, 'wishes to define, where he only investigates' and that one is generally entitled to 'expect fixed, cut-to-measure, once and for all applicable definitions in Marx's works'. He says : 'It is self-evident that where things and their interrelations are conceived, not as fixed, but as changing, their mental images, the ideas, are likewise subject to change and transformation; and they are not encapsulated in rigid definitions, but are developed in their historical or logical process of formation.'[1]

The truth of this remark can perhaps be seen best and most clearly in Marx's analysis of the conceptual distinction between fixed and circulating capital. To recapitulate : capital's principal concern in the production process was valorisation, where the only important distinction was between objectified labour and living labour. Living labour was the sole means by which capital could both maintain and increase its value. As a consequence the analysis was confined to the one crucial distinction for the valorisation of capital – that between constant and variable capital.[2]

However, valorisation only constitutes one stage in the life-span of capital. Seen as a whole capitalist production consists in the continuous alternation between its production phase and its circulation phase; it is the unity of production and circulation. 'This unity itself is movement, process', and the subject of this movement is capital – 'the value . . . predominant over the different phases of this movement, . . . sustaining and multiplying itself in it'.[3] 'The passage from

[1] *Capital* III, pp.13-14. Cf. Chapter XI of Volume II (p.230), which is headed 'Theories of Fixed and Circulating Capital' : 'It is not a question here of definitions, which things must be made to fit. We are dealing here with definite functions which must be expressed in definite categories.'

[2] 'We divided capital above into constant and variable value; this is always correct as regards capital within the production phase, i.e. in its immediate valorisation process.' (*Grundrisse*, p.649.)

[3] *ibid.* p.620.

one moment to the other appears as a particular process, but each of these processes is the transition to the other. Capital is thus posited as value-in-process, which is capital in every moment. It is thus posited as *circulating capital*;[4] in every moment capital, and circulating from one form into the next.' From this point of view 'all capital is originally circulating capital, product of circulation, as well as producing circulation . . .'[5] 'Circulating capital' is 'therefore initially not a *particular* form of capital, but is rather *capital* itself . . . as subject of the movement just described, which it, itself, is as its own valorisation process'.[6]

Nevertheless, capital is not only the unity of production and circulation, but also 'equally their *difference*, and in fact a difference distinct in space and time.' Thus if capital 'as the whole of circulation[7] is *circulating capital*, is the process of going from one phase into the other, it is at the same time, within each phase, posited in a specific aspect, restricted to a particular form, which is the negation of itself as the subject of the whole movement . . . Not-circulating capital. *Fixed* capital, actually *fixated* capital, fixated in one of the different particular aspects, phases, through which it must move.' That is as long as capital 'persists in one of these phases, the phase does not appear as fluid transition (and each of them has its duration), is not circulating [but] fixed. As long as it remains in the production process it is not capable of circulating; and it is virtually devalued. As long as it remains in circulation, it is not capable of producing, not capable of positing surplus-value, not capable of engaging in the process as capital. As long as it cannot be brought to market, it is fixated as product. As long as it has to remain on the market it is fixated as commodity. Finally, if the conditions of production remain in their form as conditions and do not enter into the production process, it is again fixated and devalued. As the subject moving through all phases, as the moving unity, the unity-in-process of circulation and production, capital is *circulating* capital; capital as restricted into any of these phases, as posited in its *divisions*, is

---

[4] Marx's original term here is *'capital circulant'*. In *Capital* II, p.156 the expression 'circling capital' is used, 'the return of the circling capital-value . . .'

[5] *Grundrisse*, p.536. Cf. *Capital* II, p.161: 'We have seen in general that all capital-value is constantly in circulation, and that in this sense all capital is circulating capital.'

[6] *Grundrisse*, p.620.

[7] 'Circulation' should be understood to mean the movement of capital through all its phases. (Cf. *Grundrisse*, p.517: 'If we now consider circulation, or the circulation of capital as a whole . . .')

fixated capital, *tied-down* capital. As circulating capital it fixates itself, and as fixated capital it circulates.' Therefore, the distinction between circulating and fixed capital 'is initially nothing more than capital itself posited in the two aspects, first as the unity of the process, then as a particular one of its phases . . .'[8] And both aspects are absolutely real (*reell*) – since capital equally represents both the unity of production and circulation, as well as their difference, and because both the continuity and the interruption of this continuity are inherent 'in the character of capital as circulating, in process'.[9]

So much on the concepts of 'circulating' (circling) and 'fixed' capital, as they emerge from the study of the total movement of capital. It is clear that the question here is not of 'two particular *kinds* of capital', but rather of '*different characteristic forms of the same capital*'.[10] '*One and the same capital* always appears in both states; this is expressed by the appearance of one part of it in one phase, another in another; one part tied down, another part circulating; circulating here, not in the sense that it is in the *circulatory phase proper* as opposed to the *production phase*, but rather in the sense that in the phase in which it finds itself, it is in a *fluid phase*, a phase-in-process, a phase in transition to the next phase; not stuck in one of them as such and hence delayed in its total process. For example : the industrialist uses only a part of the capital at his disposal . . . in production, because another part requires a certain amount of time before it comes back out of circulation. The part moving within production is then the circulating part; the part in circulation is the immobilised part . . . to be sure, sometimes one and sometimes another part is in this phase . . . but his total capital is always posited in both aspects.'

However, 'as this limit arising out of the nature of the valorisation process . . . changes with circumstances, and since capital can approach its adequate character as that which circulates, to a greater or lesser degree; since the decomposition into these two aspects . . . contradicts the tendency of capital towards maximum valorisation, it therefore invents contrivances to abbreviate the phase of fixity; and at the same time also, instead of the simultaneous coexistence of both states, *they alternate*. In one period the process appears as altogether fluid – the period of the maximum valorisation of capital; in another, a reaction to the first, the other moment asserts itself all the

---

[8] *ibid.* pp.620, 621. An echo of these arguments can be found in *Capital* II, p.47.
[9] *Grundrisse*, p.663.
[10] *ibid.* pp.621-22.

more forcibly – the period of the maximum devaluation of capital and congestion of the production process. The moments in which both aspects appear alongside one another themselves only form interludes between these violent transitions and turnings-over.' Marx notes at this juncture that 'it is extremely important to grasp these aspects of circulating and fixed capital as *specific characteristic forms* of capital generally, since a great many phenomena of the bourgeois economy – the period of the economic cycle . . ., the effect of new demand, even the effect of new gold- and silver-producing countries on general production – [would otherwise be] incomprehensible.'[11] For 'it is futile to speak of the stimulus given by Australian gold or a newly-discovered market. If it were not in the nature of capital to be never completely occupied, i.e. always partially *fixated*, devalued, unproductive, then no stimuli could drive it to greater production.'[12]

2.

However, this distinction between 'fixed' and 'circulating' capital is insufficient when we turn to the circulation process proper, the movement of capital outside the production phase. Here we see that different constituent parts of capital circulate in different ways, and therefore exhibit different turnover times. Thus the means of labour (machines etc.) never leave the actual site of production; it is only their value which circulates, through their successive and piecemeal transfer to the product. But the remaining means of production (raw material and auxiliary material),[13] and the variable capital advanced for the purchase of labour-power circulate in a quite different manner. These differing modes of circulation lead to the first factor receiving the form of 'fixed', and the second that of 'circulating' or 'fluid' capital.

Thus, whereas up to now fixed and circulating capital 'appeared to us merely as different transitory aspects of capital . . . as alternating forms of one and the same capital in the various phases of its turn-over . . . they have now hardened into two particular modes of its

[11] *ibid.* pp.622-23.
[12] *ibid.*
[13] However, 'If a means of production which is not an instrument of labour strictly speaking, e.g. an auxiliary substance, a raw material, a partly finished article, etc., behaves with regard to value-yield, and hence the manner of circulation of its value, in the same way as the instruments of labour, then it is equally a material bearer, a form of existence, of fixed capital.' (*Capital* II, p.164.)

existence', two particular kinds of capital. Insofar as 'a capital is examined in a particular branch of production, it appears as divided into these two portions, or splits into these two kinds of capital in certain proportions'.[14] 'To be fixed or circulating appears as a *particular* aspect of capital apart from that of being capital. However,' stresses Marx, 'it must proceed to this particularisation',[15] which is connected to the specific use-value of these components of capital.

The fact that we have examined the fortunes of capital in the sphere of production meant that the material differences between the various elements of production were only looked at in the context of the actual labour process, we had to differentiate between means of labour, material for labour and living labour. By contrast, in the process of the creation of value the constituent parts of capital which represent the elements of production appeared simply as quantities of value, whose only mark of distinction was the fact that one was designated as 'constant', and the other (capital laid out for the purchase of labour-power) as 'variable'. Now, however, in the categories of liquid and fixed capital, 'the relation between the factors, which had been merely quantitative . . . appears as a qualitative division within capital itself, and as a determinant of its total movement (turnover)'.[16] For a capital is only 'fixed', insofar as it physically takes on the shape of a means of labour in the production process. This implies that it gives up value to the product, and hence turns over, in a particular manner. '*The particular nature of use-value,* in which the value exists, or which now appears as capital's body, here appears as itself a *determinant* of the *form* and of the action of capital : as giving one capital a particular property as against another; as particularising it.'[17] That is, use-value reveals itself once more 'as an economic category'. However, we have already dealt with this question in more detail in Part I (in the chapter on the role of use-value in economics), and what was said there applies here as well.

[14] *Grundrisse*, p.702. Marx notes in the *Rough Draft* that 'in the human body, as with capital, the different elements are not exchanged at the same rate of reproduction, blood renews itself more rapidly than muscle, muscle than bone, which in this respect may be regarded as the fixed capital of the human body'. (*ibid.* p.670.)

[15] *ibid.* p.645. Cf. Marx's plan on p.275 of the *Grundrisse*, '(2). Particularisation of capital : (a) capital circulant, capital fixe.'

[16] *ibid.* p.692. ('The split within capital as regards its merely physical aspect has now entered into its form itself, and appears as differentiating it.' *ibid.* p.703.)

[17] *ibid.* p.646.

3.

It is unnecessary here to show in detail how the conceptual distinction between 'fixed' and 'fluid' capital is developed in the *Rough Draft*, since we encounter the results of Marx's investigation of this question in a more complete form in Volume II of *Capital*. Consequently we will confine ourselves to those points where the representation in the *Rough Draft* diverges from that in *Capital*, or where the older manuscript stresses aspects which remain in the background in *Capital* itself.

Let us look first of all at the sections superseded by the later work. According to the *Rough Draft* circulating capital consists firstly of raw materials and auxiliary materials, and secondly of the so-called *approvisionnement* of the worker i.e. his means of subsistence.[18] The latter are the object of so-called 'small-scale' circulation, as distinct from the actual or 'large-scale' circulation of capital.[19] 'This is the constantly circulating part of capital . . . which does not even for a single instant enter into its reproduction process, but constantly accompanies it . . . The worker's *approvisionnement* arises out of the production process as products, as result; but it never enters as such into the production process, because . . . it enters directly into the worker's consumption, and is directly exchanged for it. This, therefore, as distinct from raw material as well as instrument, is the circulating capital *par excellence*.'[20]

This is what the *Rough Draft* says. How is this same question answered in Volume II of *Capital*? Naturally, Marx also emphasises there that 'the money which the capitalist pays the worker for the

[18] '. . . *Approvisionnement*, as Cherbuliez calls it' refers to 'the products presupposed so that the worker lives as a worker and is capable of living during production, before a new product is created'. It is 'money expressed in the form of articles of consumption, use-values', which the workers 'obtain from the capitalist in the act of exchange between the two of them'. (*ibid.* pp.299-300.)

[19] 'Within circulation as the total process, we can distinguish between large-scale and small-scale circulation. The former spans the entire period from the moment when capital exits from the production process until it enters it again. The second is continuous and constantly proceeds simultaneously with the production process. It is the part of capital which is paid out as wages, exchanged for labour-capacity.' (*ibid.* p.673.)

[20] *ibid.* p.675. This passage concludes: 'Here is the only moment in the circulation of capital where consumption enters directly . . . Here, then — through the relation of capital to living labour-capacity and to the natural conditions of the latter's maintenance — we find circulating capital specified in respect of its use-value as well, as that which enters directly into individual consumption, to be directly used up by the latter.' (*ibid.* pp.675-76.)

use of his labour-power is nothing more or less than the form of the general equivalent for the worker's necessary means of subsistence. To this extent, the variable capital consists in substance of means of subsistence.' However, it is 'the worker himself who converts the money received for his labour-power into means of subsistence, in order to reconvert them into labour-power, to keep alive'. By contrast what the capitalist buys and consumes in the production process 'is not the worker's means of subsistence, but his labour-power itself'. 'It is therefore not the worker's means of subsistence which acquire the definite character of circulating capital as opposed to fixed capital. Nor is it his labour-power. It is rather that part of the value of productive capital which is invested in labour-power and which, by virtue of the form of its turnover, receives this character in common with some, and in contrast with other, component parts of the constant capital.'²¹ (That is, it receives this character because this part of value and similarly the value of auxiliary and raw materials, completely enters into the value of the product each time, and must therefore be completely replaced from it.)

In addition, however, *Capital* examines the reasons which led bourgeois economics to characterise the worker's means of subsistence as 'circulating' capital, in contrast to fixed capital. These originate primarily in the class nature of this school of economics – in its instinctive aversion to too deep an investigation into the 'secret of the making of profits'. 'Generally speaking, the capital advanced is converted into productive capital, i.e. it assumes the form of elements of production which are themselves the products of past labour. (Among them labour-power.) . . . Now, if instead of labour-power itself, into which the variable part of capital has been converted, we take the worker's means of subsistence, it is evident that these means as such do not differ, so far as the formation of value is concerned . . . The means of subsistence cannot themselves expand their own value or add any surplus-value to it. Their value, like that of the other elements of the productive capital, can reappear only in the value of the product. They cannot add any more value to it than they themselves possess.' Hence by characterising 'the value expended for the means of subsistence of the workers, instead of the value laid out in labour-power, as the circulating component of productive capital, the understanding of the distinction between variable and constant capital, and thus the understanding of the capitalist process of production in general, is rendered impossible. The determination that this part of capital is variable capital in contrast to the constant

²¹ *Capital* II, pp.168-69.

capital, spent for material creators of the product, is buried beneath the determination that the part of the capital invested in labour-power belongs, as far as the turnover is concerned, in the circulating part of productive capital. And the burial is brought to completion by enumerating the worker's means of subsistence instead of his labour-power as an element of productive capital.'[22]

In the *Rough Draft*, however, Marx still treated the worker's means of subsistence or the *approvisionnement* throughout as a component of circulating capital! Of course the above-mentioned explanation for this error could have played no role here, since it was precisely in the *Rough Draft* that Marx first developed the concepts of variable and constant capital, thus giving his theory of surplus-value its final shape. The source of error must therefore lie elsewhere. In our opinion it arises from the neglect of, or insufficient stress on, the perspective developed in Volume II of *Capital*; namely that in the case of the distinction between fluid and fixed capital the question is exclusively that 'of differences within the *productive* capital in the product- and value-creating process, which in turn cause differences in its turnover and reproduction'.[23] In other words, the *Rough Draft* to some extent makes the very mistake Marx later blamed Adam Smith for; namely, that he 'confuses circulating as distinguished from fixed capital with forms of capital pertaining to the sphere of circulation, with capital of circulation[24] . . . He therefore

[22] *ibid.* pp.216-18. Cf. *ibid.* pp.225-26: 'The real substance of the capital laid out in wages is labour itself, active, value-creating labour-power, living labour, which the capitalist exchanges for dead, objectified labour and embodies in his capital, by which means, and by which alone, the value in his hands turns into self-valorising value . . . But, if, on the contrary, the secondary definition of the circulating capital, which it shares with a part of the constant capital (raw material and auxiliary materials), is made the essential definition of the part of capital laid out in labour-power . . . then the part of the capital laid out in wages must likewise consist, materially, not of active labour-power but of the material elements which the worker buys with his wages, i.e. it must consist of that part of the social commodity-capital which passes into the consumption of the worker, viz., means of subsistence.'

[23] *ibid.* p.195.

[24] The term 'capital of circulation' is used in Volumes II and III of *Capital* to mean 'capital-value in those of its forms which belong in the circulation process (commodity-capital and money-capital)'. 'No matter how much money-capital and commodity-capital may function as capital and no matter how smoothly they may circulate, they cannot become circulating capital as distinct from fixed capital until they are transformed into circulating components of productive capital. But because these two forms of capital dwell in the sphere of circulation, Political Economy as we shall see has been misled since the time of Adam Smith into lumping them together with the circulating part of the productive capital . . . They are indeed capital of

mixes up commodity-capital and the circulating component of productive capital, and in that case it is a matter of course that whenever the social product assumes the form of commodities, the means of subsistence of the workers . . . must be supplied out of the commodity-capital'[25] (and from this standpoint appear to belong to 'circulating' capital).[26]

4.

Now to an aspect which is elaborated much more rigorously in the *Rough Draft* than in *Capital*, and which is related to the continually growing importance of fixed capital in the developed capitalist mode of production.[27]

The issue is that of the development of the means of labour into machinery, or into the machine system. Marx writes in the *Rough Draft* : 'As long as the means of labour remains a means of labour in the proper sense of the term, such as it is directly, historically, adopted by capital and included in its valorisation process, it undergoes a merely formal modification, by appearing now as a means of labour not only in regard to its material side, but also at the same time as a particular mode of the presence of capital, determined by its total process – as *fixed capital*.' However it does not stop at this merely formal change : 'Once adopted into the production process of capital, the instrument of labour passes through different metamorphoses, whose culmination is the *machine* or rather, an *automatic system of*

circulation in contrast to productive capital, but they are not circulating capital in contrast to fixed capital.' (*Capital* II, pp.170-71.) And not until the analysis of 'many capitals', that is, the sphere of competition, do these concepts of fixed and circulating capital receive an extended meaning, so that they can be applied to the 'fixed and circulating capital of a merchant'. (*Capital* III, 308-10.)

[25] *Capital* II, p.216.

[26] Hence the *Rough Draft* has this to say about the 'circulating products of a machine manufacturer' : 'For him they are circulating capital; for the manufacturer who uses them', (i.e. the machines), 'in the production process, fixed capital; because product for the former, and instrument of production only for the latter'. (*Grundrisse*, p.723.) Quite the reverse in *Capital*: 'In the same way a machine, the product of a machine manufacturer, is the commodity-form of his capital, is commodity-capital to him. And so long as it stays in this form it is neither circulating nor fixed capital. But if sold to a manufacturer for use it becomes a fixed component part of a productive capital.' (*Capital* II, p.210.)

[27] The passages looked at here have already been dealt with partially in Chapter 17 above.

machinery.'[28] Further, in the form of the machine, and still more in machinery as an automatic system, 'the use-value i.e. the material quality of labour, is transformed into an existence adequate to fixed capital and to capital as such; and the form in which it was adopted into the production process of capital, the direct means of labour, is superseded by a form posited by capital itself and corresponding to it.' Thus, for the first time, in machinery, 'objectified labour materially confronts living labour as a ruling power and as an active subsumption of the latter under itself, not only by appropriating it, but in the real production process itself'; and for the first time 'objectified labour appears not only in the form of product or of the product employed as means of labour, but in the form of the force of production itself . . . The accumulation of knowledge and of skill, of the general productive forces of the social brain, is thus absorbed into capital, as opposed to labour, and hence appears as an attribute of capital, and more specifically of *fixed capital*, insofar as it enters into the production process as a means of production proper. *Machinery* appears then, as the most adequate form of *fixed capital*, and fixed capital . . . as the *most adequate form of capital* as such.'[29] And it is for precisely this reason that 'the stage of development reached by the mode of production based on capital . . . is measured by the existing scope of fixed capital; not only by its quantity, but just as much by its quality'.[30]

However, as Marx goes on to say, the development of fixed capital can serve as a standard for the degree of development of capital production in yet another respect: 'The aim of production oriented directly towards use-value, as well as that oriented directly towards exchange-value, is the product itself, destined for consumption.' However, 'the part of production which is oriented towards the production of fixed capital does not produce direct objects of individual gratification, not direct exchange-values; at least not

[28] *Grundrisse*, p.692.

[29] *ibid.* pp.693-94. The situation is, however, different if we look at the reduced capacity for circulation of fixed capital. 'Precisely in this aspect as fixed capital – i.e. in the character in which capital has lost its fluidity and become identified with a specific use-value, which robs it of its ability to transform itself – does developed capital . . . most strikingly manifest itself.' However, from this standpoint, fixed capital does not correspond to the concept of capital, 'which, as value, is indifferent to every specific form of use-value, and can adopt or shed any of them as equivalent incarnations', so that in this respect 'it is circulating capital which appears as the adequate form of capital, and not fixed capital'. Marx adds: 'This contradiction pretty. To be developed.' (*ibid.* pp.679, 694.)

[30] *ibid.* p.715.

directly realisable exchange-values. *Hence, only when a certain degree of productivity has already been reached – so that a part of production-time is sufficient for immediate production – can an increasingly large part be applied to the production of the means of production.* This requires that society be able to wait; that a large part of the wealth already created can be withdrawn both from immediate consumption and from production for immediate consumption, in order to employ this part for labour which is *not immediately productive* (within the material production process itself) . . . As the *magnitude of relative surplus labour depends on the productivity of necessary labour, so does the magnitude of labour-time* – living as well as objectified – *employed on the production of fixed capital depend on the productivity of the labour-time spent in the direct production of products.*[31] *Surplus population* (from this standpoint),[32] as well as *surplus production,* is a condition for this. That is, the output of the time employed in direct production must be larger, relatively, than is directly required for the reproduction of the capital employed in these branches of industry. The *smaller* the direct fruits borne by *fixed capital,* the less it intervenes in the *direct production process,* the greater must be this relative *surplus population and surplus production*; thus, more to build railways, canals, aqueducts, telegraphs etc. than to build the machinery directly active in the direct production process.'[33]

And, in another passage : 'Insofar as the production of fixed capital, even in its physical aspect, is not directed immediately towards the production of values required for the direct reproduction of capital i.e. those which themselves in turn represent use-value in the value-creation process – but rather towards the production of the means of value-creation . . . (the production of value posited physically in the object of production itself, as the aim of production . . .), it is in the production of *fixed capital that capital posits itself as end-in-itself,* and appears active as capital, *to a higher power than it does in the production of circulating capital.* Hence, in this respect as well, the dimension already possessed by fixed capital, which its production occupies within total production, is the measur-

[31] 'The labour-time employed in the production of fixed capital relates to that employed in the production of circulating capital, within the production process of capital itself, as does surplus labour-time to necessary labour-time.' (*ibid.* p.709.)

[32] That is, not in the sense of the 'industrial reserve army'.

[33] *Grundrisse,* p.707.

ing rod of the *development* of wealth founded on the mode of production of capital.'[34]

These passages are certainly a valuable complement to Volume II of Marx's *Capital*. But in fact, the *Rough Draft* goes even further; it also draws a picture of a society where the development of machinery and of the general conditions of production [35] is taken so far that it is no longer the 'direct labour which man himself performs, nor the time which he works, but rather the appropriation of his own general productive power . . . which appears as the great foundation of production and wealth'; a society, therefore, where the law of value itself must disappear. However, we shall leave this aspect until later (the chapter on 'the historical barrier to the law of value').

5.

Up till now we have discussed the changes which the capitalist production process undergoes owing to the development of fixed capital. However, what is the impact of this development – i.e. the constant growth and growing significance of values confined to the form of machinery – on the circulation process of capital?

The general answer runs as follows : 'With circulating capital, reproduction is determined by circulation time; with fixed capital, circulation is determined by the time in which it is consumed as use-value, in its material presence, within the act of production, i.e. by the period of time within which it must be reproduced.'[36] However, this distinction means that 'the *turnover time* of a total capital divided into circulating and fixed capital becomes essentially modified'.[37] For example : if a capital consists of £10,000, of which £5,000 is fixed, and £5,000 circulating (to use a calculation from the *Rough Draft*), and if the latter turns over once a year while the former turns over once every five years, then, 'in 20 months the total capital of £10,000 is turned over, although the fixed capital is replaced only in 5 years. This turnover time holds, however, only for the repetition of the production process . . . not for the reproduction of the capital itself'.[38] This is because the capital itself is not, needless to say,

---

[34] *ibid.* p.710.
[35] These 'general' or 'communal' conditions of production are understood to be canals, railways etc. in the *Rough Draft*.
[36] *ibid.* p.682.
[37] *ibid.*
[38] *ibid.* p.718.

replaced by the average turnover,[39] and not until five years have passed will the capitalist in fact 'be once more in possession of the total capital with which he began the production process'. Thus although '*in the creation of surplus-value his capital acted as if it had wholly turned over in 20 months . . . the total capital itself is only reproduced in 5 years. The former aspect of turnover important for the proportion in which it valorises itself*; the latter however brings in a new relation, which does not hold at all with circulating capital.' Namely because 'circulating capital is completely absorbed into circulation and returns from it as a whole, it follows that it is reproduced as capital as many times as it is realised as surplus-value, or as surplus capital. But since fixed capital never enters circulation as a use-value, and enters it as value only to the extent that it is consumed as a use-value, it follows that it is by no means reproduced as soon as the *surplus-value determined by the average turnover time of the total capital is posited.* The turnover of the circulating capital must take place 10 times in the 5 years before the fixed capital is reproduced; i.e. the period of the revulsions of circulating capital must be repeated 10 times while that of fixed capital is repeated once, and the *total average turnover of the capital – 20 months – has to be repeated 3 times before the fixed capital is reproduced.*[40] Hence, the larger the part of the capital consisting of fixed capital – i.e. the more capital acts in the mode of production corresponding to it, with great employment of produced productive force – and the more durable the fixed capital i.e. the longer its reproduction time, the more its use-value corresponds to its specific economic role – the more often must the part of capital which is determined as circulating *repeat the period of its turnover, and the longer is the total time the capital requires for the achievement of its total circulation.* Hence the *continuity* of production becomes an external necessity for capital with the development of that portion of it which is determined as fixed capital. For circulating capital, *an interruption, if it does not last so*

---

[39] This is evident with circulating capital: 'If a capital of 100 returns 4 times a year and hence brings in 20 per cent, like a capital of 400 which circulates only once, then the capital remains 100 at the end of the year as at the beginning, and the other capital remains 400, although it has effected a production of use-values and a positing of surplus-value equal to a 4 times larger capital. The fact that the velocity of turnover here substitutes for the magnitude of the capital shows strikingly that it is only the amount of surplus labour set into motion, and of labour generally which determines the creation of value as well as the creation of surplus-value, and not the magnitude of the capital for itself.' (*ibid.* p.718.)

[40] Cf. *Capital* II, pp.185-87.

long as to ruin its use-value, is only an interruption in the creation of surplus-value. But with fixed capital, the interruption, insofar as in the meantime its use-value is necessarily destroyed relatively unproductively, i.e. without replacing itself as value, is the destruction of its original value itself. Hence the continuity of the production process which corresponds to the concept of capital is posited as *conditio sine qua* [*non*] for its maintenance only with the development of fixed capital; hence likewise the continuity and the constant growth of consumption.'[41]

But this is not all. Marx says that the second result to which the examination of the influence of fixed capital on turnover time leads 'is from the formal side, even more important'. 'The total time in which we measured the return of capital was the year, while the time unit in which we measure labour is the day. We did so firstly because the year is more or less the natural reproduction time, or duration of the production phase, for the reproduction of the largest part of the vegetable raw materials used in industry. The turnover of circulating capital was determined, therefore, by the number of turnovers in the total time of a year.'[42] But in reality 'the circulating capital begins its reproduction at the end of each turnover, and while the number of turnovers during the year affects the total value, and the fate it encounters during each turnover appears a determinant of the conditions under which it begins reproduction anew, each of them for itself is, nevertheless, a complete life-span for the circulating capital. As soon as capital is transformed back into money, it can, e.g. . . . throw itself from one branch of production into another, so that *reproduction, regarded materially, is not repeated in the same form.*'

'The introduction of fixed capital,' Marx goes on to say,

---

[41] *Grundrisse*, p.719.

[42] 'Considering that the production process of capital is at the same time a technological process – production process absolutely – namely [the process] of the production of specific use-values through specific labour, in short, in a manner determined by this aim itself; considering that the most fundamental of these processes is that through which the body reproduces its necessary metabolism, i.e. creates the necessaries of life in the physiological sense : considering that this production process coincides with agriculture; and the latter also at the same time directly (as with cotton, flax etc.) furnishes a large part of the raw materials for industry (actually all except those belonging to the extractive industries); considering that reproduction in agriculture in the temperate zone (the home of capital) is bound up with the general terrestrial circulation; i.e. harvests are mostly annual; it follows that the year (except that it is reckoned differently for various productions) has been adopted as the general period of time by which the sum of the turnovers of capital is calculated and measured.' (*ibid.* pp.639-40.)

'changes this; and neither the turnover time' of the circulating capital, 'nor the unit in which the number of times it turns over is measured, the year, henceforth appear as the measure of time for the motion of capital. This unit is now determined, rather, by the *reproduction* time required for fixed capital, and hence the total circulation time it needs to enter into circulation as value, and to come back from it in the totality of its value. The reproduction of the circulating capital must *also proceed in the same material form* during this whole time, and the number of its necessary turnovers ... *is distributed over a longer or shorter series of years.* Hence a longer *total period* is posited as the unit in which its turnovers are measured, and their *repetition* is now not merely externally, but rather necessarily connected with this unit.'[43] (Marx assumes in the *Rough Draft* that we are dealing with a period of approximately ten years.[44]) The importance of this becomes evident from the fact that 'the cycle which industry has passed through since the development of fixed capital on a large scale, at more or less 10-yearly intervals, is connected with this *total reproduction phase of capital*',[45] so that the average time in which machinery is renewed represents a basis for determining the periodicity of crises.[46] This is a line of reasoning which we find, further developed, in Volume II of *Capital*.[47]

[43] *Grundrisse*, p.720.
[44] Cf. Marx's discussion with Engels on this point in *MEW* Vol.29, pp.291-93. (Letters of 2 March 1858 and 4 March 1858.)
[45] *Grundrisse*, p.720.
[46] 'We shall find other determinant causes as well. But this is one of them. There were good and bad times for industry too, as well as for harvests (agriculture). But the industrial cycle of a number of years, divided into characteristic periods, epochs, is peculiar to large-scale industry.' (*ibid.* pp.720-21.)
[47] *Capital* II, pp.188-189.

PART FIVE
Capital as Fructiferous. Profit and Interest

25.
The Transformation of Surplus-Value into Profit. The General Rate of Profit

As the title indicates, the last section of the *Rough Draft* (pp. 745 ff) corresponds in a certain sense to Volume III of *Capital*. However, this is only in a certain sense, since, apart from its sketchy nature,[1] the categories of profit and interest are only examined in this section insofar as they are taken from the analysis of 'capital in general'. This constitutes the key difference between this section and Parts 1-3 of Volume III of *Capital*.[2]

The section on profit and interest begins with the following passage which is very Hegelian in flavour : 'Capital is now posited as the unity of production and circulation' (i.e. after it has described its complete circuit) '. . . realised not only as value which reproduces itself and is hence perennial, but also as value which posits value. Through the absorption of living labour-time and through the move-

---

[1] Out of the entire section of over 130 pages, 40 at most are devoted to profit and interest. The remaining pages deal with the history of the theory of money, the 'recapitulation' of theories of surplus-value etc. In fact we should also take the preceding section into consideration, since it contains a number of discussions which belong to Section Three. (Marx says on this: 'A very large part of what belongs here has been developed above. But the anticipated material is to be put here.')

[2] Cf. pp.13-14 above.

ment of its own circulation (in which the movement of exchange is posited as its own, as the inherent process of objectified labour), it relates to itself as positing new value, as producer of value. It relates as the foundation to surplus-value as that which it founded[3]. . . . In a definite period of time which is posited as the unit measure of its turnovers . . . capital produces a definite surplus-value, which is determined not only by the surplus-value it posits in one production process, but rather by the number of repetitions of the production process, or of its reproduction in a specified period of time. Because of the inclusion of circulation, of its movement outside the immediate production process, within the reproduction process, surplus-value *appears*[4] no longer to be posited by its [i.e. capital's] simple direct relation to living labour; this relation appears, rather, as merely a moment of its total movement. Capital . . . therefore no longer measures the newly produced value by its real measure, the relation of surplus labour to necessary labour, but rather by itself as its pre-supposition. A capital of a certain value produces, in a certain period of time, a certain surplus-value. Surplus-value is thus measured by the value of the presupposed capital, capital thus posited as self-valorising value – is *profit* . . . and the *rate of profit* is therefore determined by the proportion between its value and the value of capital.'[5]

At first sight this may appear to be a contrived *a priori* construction. In fact this is the point where Marx first begins to expound that same line of thought which we find in a much more developed form in *Capital* (and in the *Theories*), and which forms the basis of his theory of profit. Namely, that the category of profit should not be confused with that of *surplus-value* (as the classical economists did).[6] Rather, profit must be understood as a '*secondary*, derivative . . . form, developed further in the sense of capital . . . the bourgeois

---

[3] There is a very similar formulation in *Capital*: the advanced money capital 'is capital by virtue of its relation to the other part of M'' – the valorised capital – 'which it has brought about, which has been effected by it as the cause, which is the consequence of it as the ground'. (*Capital* II, p.45.) See note 107 on p.37 above.

[4] 'The essence must appear.' Hegel, *Science of Logic* II, p.107. Cf. *Capital* I, p.682 (542): 'The form of appearance . . . as contrasted with the essential relation manifested in it.'

[5] *Grundrisse*, pp.745-46.

[6] In fact, at the beginning of the *Rough Draft* (in the section on the production process, pp.342-44) the expressions 'rate of profit' and 'rate of surplus-value' are not strictly separated and are even identified.

form, in which the traces of its origin are extinguished'.[7] And what Marx had to say on the subject of 'all forms of appearance and their hidden background' also applies in this case. 'The forms of appearance are reproduced directly and spontaneously as current and usual modes of thought; the essential relation must first be *discovered* by science.'[8]

In fact, profit 'in its immediate form . . . is nothing but the sum of the surplus-value expressed as a proportion of the total value of the capital'.[9] It follows from this, then, (1) that the total sum of profit (of the capitalist class),[10] can never be greater than the total sum of surplus-value, and (2) that – regarded as the rate of profit – profit must '*under all circumstances . . . express a smaller proportion of the gain than the real proportion of the surplus-value.* For under all circumstances it is measured by the total capital, which is always

---

[7] *Grundrisse*, pp.595, 762. We should not overlook the fact that the subsequent transformation of surplus-value into the form of profit simply represents 'a further development of the inversion of subject and object that takes place already in the process of production'. Marx states that there we have already seen that 'the subjective productive forces of labour appear as productive forces of capital. On the one hand, the value, or the past labour, which dominates living labour, is incarnated in the capitalist. On the other hand, the worker appears as bare material labour-power, as a commodity.' And precisely because 'at one pole the price of labour-power assumes the transmuted form of wages, surplus-value appears at the opposite pole in the transmuted form of profit'. (*Capital* III, pp.37, 45.) Cf. Marx's letter to Engels of 30 April 1868: 'As, owing to the form of wages, the whole of labour appears to be paid for, the unpaid part of labour seems necessarily to come not from labour but from capital, and not from the variable part of capital but from capital as a whole. In this way surplus-value assumes the form of profit.' (*Selected Correspondence*, p.192.)

[8] *Capital* I, p.682 (542).

[9] *Grundrisse*, p.767. The course of the analysis will show how – as a result of the formation of a general rate of profit – 'the alienation goes further, and how profit represents a magnitude differing also numerically from surplus-value'. (*Capital* III, p.48.) Cf. *Theories* III, pp.482-83: 'Furthermore, as a result of the conversion of profit into average profit, the establishment of the general rate of profit and, in connection with it and determined by it, the conversion of values into cost-prices, the profit of the individual capital becomes different from the surplus-value produced by the individual capital in its particular sphere of production, and different, moreover, not only in the way it is expressed – i.e. rate of profit as distinct from rate of surplus-value – but it becomes substantially different, that is, in this context, quantitatively different. Profit does not merely seem to be different, but is now in fact different from surplus-value, not only with regard to the individual capital, but also with regard to the total capital in a particular sphere of production.'

[10] 'Profit as we still regard it here, i.e. as the profit of capital as such, not of an individual capital at the expense of another, but rather as the profit of the capitalist class.' (*Grundrisse*, p.767.)

larger than that employed for wages and exchanged for living labour.'[11] Consequently, the rate of profit 'never expresses the real rate at which capital exploits labour, but always a much smaller relation'. It could 'express the real rate of surplus-value only if the entire capital were exchanged for living labour . . . hence if not only the raw material were = 0, but also the means of production'. However, the latter 'cannot occur on the basis of the mode of production corresponding to capital'.[12]

Thus, since from the outset the rate of profit (as distinct from profit as such) differs qualitatively from the rate of surplus-value, the laws of its movement do not coincide 'so directly or simply' with those of the rate of surplus-value as might appear initially.[13] 'The rate of profit can fall, although real surplus-value increases. The rate of profit can rise although real surplus-value falls.' This already follows from the fact that the rate of profit is calculated on the total value of the capital. It is therefore 'determined (1) by the magnitude of the surplus-value itself; (2) by the relation of living labour to accumulated.'[14] (i.e. by the value composition of capital.) And finally, differences in turnover time also affect the size of the surplus-value produced, and therefore the rate of profit.

This leads to the result that the same rate of profit can in fact be based on very different rates of surplus-value, and conversely, 'one and the same rate of surplus-value may be expressed in the most varying rates of profit'.[15] Thus the degree of exploitation of labour can be the same in different branches of production, with the rate of surplus-value at the same level : however, since the organic composition of capital varies from branch to branch, these branches will produce very different masses of surplus-value, and these masses will be expressed in widely varying rates of profit.[16] Indeed it is precisely 'the inequality of profit in different branches of industry with capitals of equal magnitudes [which] is the condition and presupposition for their equalisation through competition'.[17]

We thus arrive at the problem of the general rate of profit, and

---

[11] *ibid.* pp.767, 753.
[12] *ibid.* pp.762-63.
[13] *Theories* II, p.426.
[14] *Grundrisse*, pp.747, 817.
[15] *Capital* III, p.68.
[16] On the other hand : 'If capitals whose component parts are in different relations, including therefore their forces of production, nevertheless yield the same percentages on total capital, then the real surplus-value has to be very different in the different branches.' (*Grundrisse*, p.395.)
[17] *ibid.* p.761.

prices of production which diverge from values. We are reminded of Böhm-Bawerk's assertion that there is an 'irreconcilable contradiction' between the first and third Volumes of *Capital*, and that the theory of average profit developed in Volume III is to be understood as basically a retreat 'under fire', as an 'act of self-defence in anticipation'. Hilferding's reply to this was that the relevant section of Volume III was actually written in 1865, i.e. two years before the publication of Volume I. We shall now see that the problem of the average rate of profit was in fact already solved in the *Rough Draft* of 1857-58, i.e. before Marx had even set out his theory of value! We read in the *Rough Draft* : 'Since the profit of capital is realised only in the price which is paid for it, for the use-value created by it, profit is determined by the *excess of the price obtained over the price which covers outlays*' i.e. over 'cost price'.[18] 'Since, furthermore, this realisation proceeds only through *exchange*, the individual capital's *profit is not necessarily restricted by its surplus-value*, by the surplus labour contained in it : but is relative, rather, to the excess of price obtained in exchange. It can exchange more than its *equivalent, and then its profit is greater than its surplus-value*. This can be the case only to the extent that the other party to the exchange does not obtain an equivalent.' On the other hand, profit can also be smaller than surplus-value, i.e. 'it can exist for capital, even without the realisation of the real production costs – i.e. the whole surplus labour set to work by capital'. However, 'the total surplus-value, as well as the *total profit*, which is only *surplus-value itself, computed differently*, can neither grow nor decrease through this operation, ever; what is *modified thereby* is not it, but only *its distribution among the different capitals*.'[19]

How does this distribution take place? The answer is provided in an excursus in the section of the *Rough Draft* dealing with the

[18] 'In relation to profit, the value of the capital presupposed in production appears as advances – production costs which must be replaced in the product. After deduction of the part of the price which replaces them, the excess forms the profit. Since surplus labour . . . costs capital nothing, hence does not figure as part of the value advanced by it . . . it follows that this surplus labour, which is included in the production costs of the product and forms the source of surplus-value and hence of profit as well, does not figure as part of the production costs of capital. The latter are equal only to the values actually advanced by it, not including the surplus-value appropriated in production and realised in circulation. The production costs from the standpoint of capital are therefore not the real production costs, precisely because surplus labour does not cost it anything. The excess of the price of the product over the price of the production costs gives it its profit.' (*Grundrisse*, p.760.)

[19] *ibid.*

circulation process : 'A *general rate of profit* as such is possible only if the rate of profit in one branch of business is too high and in another too low; i.e. if a part of the surplus-value – which corresponds to surplus labour – is transferred from one capitalist to the other. If, in 5 branches of business, for example, the rate of profit is respectively

A	B	C	D	E
15%	12%	10%	8%	5%

then the average rate is 10%; but, in order for this to exist in reality, capitalist A and B have to give up 7% to D and E – more particularly 2 to D and 5 to E – while C remains as it was. It is impossible for rates of profit on the same capital of 100 to be equal, since the relations of surplus labour are altogether different, depending on the productivity of labour and on the relation between raw material, machinery and wages and on the overall volume in which production takes place ... The capitalist class, thus, to a certain extent distributes the total surplus-value so that' the capitalists participate in it 'evenly, in accordance with the *size* of their capital, instead of in accordance with the surplus-values actually created by the capitals in the various branches of business. The larger profit – arising from the real surplus labour within a branch of production, the really created surplus-value – is pushed down to the average level by competition', while 'the deficit of surplus-value in the other branch of business is raised up to the average level by withdrawal of capitals from it ... This is realised by means of the relation of prices in the different branches of business, which fall *below* their *value* in some, rise *above* it in others.[20] This makes it seem as if an equal sum of capital in unequal branches of business created *equal surplus labour or surplus-value.*'[21] Marx remarks, however, that this question belongs first of all 'in the section on competition', 'of many capitals, not here',[22] where we are only concerned with '*the* profit of *capital*' (i.e. with capital and profit 'in general').[23]

Marx adds : 'It is altogether necessary to make this clear; because the distribution of the surplus-value among the capitals . . .

[20] In this sense Marx already speaks in the Rough Draft of 'price as market price or the general price'. The expression 'price of production (*Produktionspreis*) first appears in the *Theories*. (Cf. on this Kautsky's note on pp.15-16 of Vol.II of his edition of the *Theories* and Marx's letter to Engels of 2 August 1862. *Selected Correspondence*, pp.120-23, where it is referred to as 'cost price'.)
[21] *Grundrisse*, pp.435-36.
[22] *ibid.* pp.435, 760.
[23] *ibid.* p.787.

this *secondary* economic operation – gives rise to phenomena which are confused with the primary ones.' ('It is clear that other aspects also enter in with the equalisation of the rate of profit. Here, however, the issue is not the distribution of surplus-value but its creation.'[24]) However, both levels of the analysis are necessary since 'the greatest confusion and mystification has arisen because the doctrine of surplus profit has not been examined in its pure form by previous economists, but rather mixed in together with the doctrine of real profit, which leads up to distribution, where the various capitals participate in the general rate of profit'.[25] Thus the case of Ricardo, whose theory of profit cannot overcome the contradiction between the determination of the values of products by relative labour-time and the *'real determination of prices, in practice'* precisely because he 'does not grasp profit as itself a derivative, secondary form of *surplus-value'*.[26]

This leads us to the question of the relation of Marx's theory of profit to Ricardo's (and that of the classical economists in general). The difference between the two theories is immediately apparent.

[24] *ibid.* pp.632, 669.
[25] *ibid.* p.684. We should refer above all here to the illusion arising out of the 'division of surplus-value into average portions', according to which 'all parts of capital equally bring a profit'. Of course, 'if I take the total value of the finished product, then I can compare every part of the product advanced with the part of the outlay corresponding to it; and the percentage of profit in relation to the whole product is naturally the same percentage for any fractional part of the product . . . This obviously means nothing other than that if I gain 10 per cent on 100 then the gain on every part of 100 amounts to as much as, when added together, will be 10 per cent on the total sum.' But 'it is impossible to see what use this calculation is'. (*ibid.* pp.723, 567-68.) This illusion seems to have been taken to the absurd in the case of the 'marvellous invention of Dr. Price' (1772), according to which, 'One penny, put out at our Saviour's birth to 5 per cent compound interest, would, before this time, have increased to a greater sum, than would be contained in a hundred and fifty millions of earths, all solid gold.' Price was misled into this fantasy because 'he took no note of the conditions of reproduction and labour, and regarded capital as a self-regulating automaton, as a mere number that increased itself'. However, 'the identity of surplus-value and surplus labour imposes a qualitative limit upon the accumulation of capital. This consists of the total working day, and the prevailing development of the productive forces and of the population, which limits the number of simultaneously exploitable working days. But if one conceives of surplus-value in the meaningless form of interest, the limit is merely quantitative and defies all fantasy . . . Practice has shown the economists that Price's interest-multiplication is impossible; but they have never discovered the blunder contained in it.' (*Capital* III, pp.394-95, 398-99.) Except for the last sentence this passage was taken over, with only slight stylistic changes, from the *Rough Draft*, pp.375, 842-43.
[26] *Grundrisse*, p.554.

Whereas the Ricardian school came to grief on the contradiction between the determination of value by labour and the existence of the general rate of profit, this contradiction provided the point of departure for Marx's new theory of profit. Unlike the Ricardians, he does not attempt to rescue the law of value 'from the contradictions of immediate experience by making a violent *abstraction*',[27] but demonstrates, on the contrary, how, by means of the intervention of the general rate of profit, 'a market price differing from this exchange-value comes into being . . . on the basis of exchange-value . . . or more correctly, how the law of exchange-value is realised only through its own antithesis'.[28] One can therefore understand the satisfaction which Marx expressed about this particular achievement of his theory in a letter to Engels on 14 January 1858. He writes, 'I am getting some nice developments. e.g. I have overthrown the entire doctrine of profit as previously conceived. In the *method* of working it was of great service to me that by mere accident I leafed through Hegel's *Logic* again.' And he added : 'If there should ever be a time for such work again, I should very much like to make accessible to the ordinary human intelligence – in two or three printer's sheets – what is *rational* in the method which Hegel discovered but at the same time enveloped in mysticism.'[29]

We now know what the 'overthrowing' of previous theories of profit consisted in : namely, the scientific understanding of profit as a 'necessary form of appearance' of surplus-value. But not only that. Marx's solution to the problem of the general rate of profit required many intermediate links; it not only presupposed a theory of production prices and cost prices, but also a correct understanding of the turnover of capital, and above all, of the problem of surplus-value. On the other hand an elucidation of the problem of surplus-value was not possible, so long as the fundamental distinction between variable and constant capital remained unrecognised, which in turn presupposed the discovery of the dual character of the labour contained in commodities. All these intermediate links are absent in Ricardo and the other classical economists. It is no surprise, then, that Ricardo 'seeks *directly* to prove the congruence of the economic categories with one another'[30] and 'arbitrarily' to equalise the

---

[27] *Capital* I, p.421 (307).

[28] *Contribution*, p.62.

[29] *Selected Correspondence*, p.93. We have been able to confirm many times in the course of this work that 'leafing through' Hegel's *Logic* not only contributed to the solution of the problem of profit, but also many others.

[30] *Theories* II, p.165. ('He never analysed the form of the mediation.' *Grundrisse*, p.327.)

rate of profit with the rate of surplus-value.[31] Hence his attempt 'to derive undeniable empirical phenomena by simple formal abstraction directly from the general law . . . The vulgar mob has therefore concluded that theoretical truths are abstractions which are at variance with reality, instead of seeing, on the contrary, that Ricardo does not carry true abstract thinking far enough, and is therefore driven into false abstraction.'[32] In other words : Ricardo lacks the dialectical incisiveness which is required to understand capital as a 'unity-in-process' and elaborate its contradictions. The chief defect of the Ricardian theory of profit was therefore its inadequate method – and this was the pivot which Marx could use to 'overthrow' this theory. In this respect the service rendered by Hegel's *Logic* cannot be rated highly enough.[33]

[31] *Theories* III, p.338.
[32] *Theories* I, p.89, II, p.437.
[33] As one critic of Marx has rightly said : 'His basic philosophical position is evident through all the fissures in his system. He approaches the object of his study, bourgeois society, with Hegelian methods, Hegelian modes of thought and Hegelian concepts.' (E.Preiser, *Das Wesen der Marxschen Krisentheorie*, p.272.)

26.

The Law of the Falling Rate of Profit and the Tendency of Capitalism Toward Breakdown

The manuscript of 1857-58 also offers the solution to yet another fundamental question in economics; that of the tendency of the rate of profit to fall.

This solution also emerged in the course of Marx's confrontation with Ricardo who, like all the classical economists, stressed the fact that with the accumulation of capital 'the natural tendency of profit is to fall'.[1] But what is the source and basis of this tendency?

Ricardo was clearly unsatisfied by Adam Smith's explanation. We read in the *Rough Draft* : 'A.Smith explained the fall of the rate of profit, as capital grows, by the competition among capitals. To which Ricardo replied that competition can indeed reduce profits in the various branches of business to an average level, can equalise the rate, but cannot depress the average rate itself.' Marx continues : 'A. Smith's phrase is correct to the extent that only in competition – the action of capital upon capital – are the inherent laws of capital, its tendencies, realised.[2] But it is false in the sense in which he understands it, as if competition imposed laws on capital from the outside, laws not its own. Competition can permanently depress the rate of profit, if . . . and insofar as a general and permanent fall of the rate of profit, having the force of a law, is conceivable *prior* to competition and regardless of competition.' 'To try to explain the inner laws of capital simply as results of competition means to concede that one does not understand them.'[3] However, what is the inner law according to Ricardo himself, which produces the tendency of the rate of profit to fall?

Let us first recall that Ricardo was unaware both of the distinction between constant and variable capital[4] and between the rate

---

[1] Ricardo, *Principles of Political Economy and Taxation*, p.139.

[2] Cf. pp.115ff above.

[3] *Grundrisse*, pp.751-52.

[4] For this reason also 'he nowhere touches on or perceives the differences in organic composition within the actual process of production'. (*Theories* II, p.373.)

of profit and the rate of surplus-value. Furthermore, according to his theory, profits and wages could only rise or fall in an inverse relation to one another. Hence his thesis that 'no accumulation of capital will permanently lower profits unless there be some permanent cause for the rise of wages'.[5] However, what are the conditions under which the wage (which for Ricardo usually remains equal to the price of the workers' necessary means of subsistence) would continually rise in terms of value (not use-value) – so that the part of the working day which the worker works for himself would grow, but the part he gives *gratis* to the capitalist would get smaller? This is clearly only possible 'if the value of the means of subsistence, on which the worker spends his wage, increases. But as a result of the development of the productivity of labour, the value of industrial commodities is constantly decreasing. The diminishing rate of profit can therefore only be explained by the fact that the value of food, the principal component part of the means of subsistence, is constantly rising.'[6] According to Ricardo this happens 'because agriculture is becoming less productive ... The continuous fall in profit is thus bound up with the continuous rise in the rate of rent.'[7]

What is evident from this is that Ricardo's explanation of the law of the fall in the rate of profit is based on two presuppositions : 1. the Malthusian presupposition of the declining fertility of agriculture, of the progressive deterioration of the soil under cultivation; and 2. the 'false assumption that the rate of profit is equal to the rate of relative surplus-value[8] and can only rise or fall in inverse proportion to a fall or rise in wages'.[9]

As is well known, Marx rejected Ricardo's solution to the problem, although we cannot deal here with the numerous reasons

[5] Ricardo, *op. cit.*, p.290. (Cf. *Theories* II, pp.466-68.)

[6] Cf. Ricardo, *op. cit.*, p.139 : 'the theory, that profits depend on high or low wages, wages on the price of necessaries, and the price of necessaries chiefly on the price of food, because all other requisites may be increased almost without limit'. Cf. also p.296 : 'it may be added that the only adequate and permanent cause for the rise of wages is the increasing difficulty of providing food and necessaries for the increasing number of workmen'.

[7] *Theories* II, pp.438-39. 'The falling rate of profit hence corresponds, with him, to the nominal growth of wages and real growth of ground-rent.' (*Grundrisse*, p.752.)

[8] Marx uses the expression relative surplus-value here because Ricardo 'assumes the working day to be constant', and thus only considers changes in relative surplus-value. (*Theories* II, p.439.)

[9] *ibid.* p.439.

he put forward.[10] The sole issue for us in this context is that Ricardo's incorrect theory of profit prevented him from explaining 'one of the most striking phenomena of capitalist production' – the tendency of the rate of profit to fall.[11] 'Since Ricardo simply mixes surplus-value and profit together in this way, and since the surplus-value can constantly decline, can *tendentially* decline only if the relation of surplus labour to necessary labour, i.e. to the labour required for the reproduction of labour-capacity, declines, but since the latter is possible only if the productive force of labour declines, Ricardo assumes that the productive force of labour decreases in agriculture, although it grows in industry, with the accumulation of capital. He flees from economics to seek refuge in organic chemistry.'[12]

But how did Marx solve this problem? In Section One of the *Rough Draft* he asks, in relation to one of the numerical examples, with which he wants to explain the distinction between the rate of profit and the rate of surplus-value : 'But, understood differently, is there not after all something correct in these figures?' Is it not possible for surplus-value, to 'rise, although in relation to the capital as a whole it declines i.e. the so-called rate of profit declines?'[13] 'The solution of the whole matter', says Marx later in a direct polemic against Ricardo, 'is simply that the rate of profit is not the same as the absolute surplus-value, but is rather the surplus-value in relation to the capital employed, and that the growth of productive force is accompanied by the decrease of that part of capital which represents *approvisionnement*[14] in relation to that part which represents invariable capital' i.e. constant capital;[15] 'hence, when the relation between total labour and the capital which employs it falls, then the part of labour which appears as surplus labour, or surplus-value, necessarily falls too'.[16] In other words; since the rate of profit is in no way identical with the rate of surplus-value, the decrease of variable capital in relation to constant brought about by the continual revolution in the techniques of production, and the increase in productivity, must express itself in a declining rate of profit. (This is a con-

---

[10] These can be found in the *Grundrisse*, pp.333, 385-86, 557-58, 596, 751-54, 756-57; in *Theories* II, pp.438-39, 463-64, 467-68, 541-46; *Theories* III, 106-09, 351-52; as well as in *Capital* III, pp.279-80.
[11] *Grundrisse*, p.558.
[12] *ibid.* pp.753-54.
[13] *ibid.* pp.380, 381.
[14] See p.356 above.
[15] See p.358 above on the initial variations in the *Rough Draft* in the use of the terms 'constant' and 'variable' capital.
[16] *Grundrisse*, p.558.

clusion which, as Marx pointed out in his letter to Engels of 30 April 1868, follows directly from the 'law of the increasing growth of the constant part of capital in relation to the variable' i.e. the increasing organic composition of capital elaborated in the presentation of the production process.)[17] 'The growth of the productive power of labour is identical in meaning with (a) the growth of relative surplus-value or of the relative surplus labour-time which the worker gives to capital; (b) the decline of the labour-time necessary for the reproduction of labour-capacity; (c) the decline of the part of capital which exchanges at all for living labour, relative to the parts of it which participate in the production process as objectified labour and as presupposed value. The profit rate is therefore inversely related to the growth of relative surplus-value or of relative surplus labour, to the development of the powers of production, and to the magnitude of the capital employed as constant capital within production.'[18] 'Thus, in the same proportion as capital takes up a larger place as capital in the production process relative to immediate labour, i.e. the more the relative surplus-value grows – the value-creating power of capital – *the more does the rate of profit fall.*'[19] In reality, however, this fall in the rate of profit occurs 'only as a tendency, like all other economic laws',[20] and is checked by numerous 'countervailing influences'. We read in the *Rough Draft*: 'There are moments in the developed movement of capital which delay this movement' i.e. the fall in the rate of profit 'other than by crises; such as e.g. the constant devaluation of a part of the existing capital : the transformation of a great part of capital into fixed capital which does not serve as agency of direct production; unproductive waste of a great portion of capital etc. . . . The fall [in the rate of profit] – likewise delayed by the creation of new branches of production in which more direct labour in relation to capital is needed, or where the productive power of labour is not yet developed . . . (similarly monopolies) . . . That the fall in the rate of profit can further be delayed by the omission of existing deductions from profit, e.g. by a lowering of taxes, reduction of ground-rent etc., is actually not our concern here, although of importance in practice, for these are themselves portions of the profit

[17] *Selected Correspondence*, pp.191-95.
[18] *Grundrisse*, p.763.
[19] *ibid.* p.747.
[20] *Capital* III, p.175. Cf. (*ibid.*): 'But in theory it is assumed that the laws of capitalist production operate in their pure form. In reality there exists only approximation; but this approximation is the greater, the more developed the capitalist mode of production and the less it is adulterated and amalgamated with survivals of former economic conditions.'

under another name, and are appropriated by persons other than the capitalists themselves.'[21]

A closer examination would show that the factors which delay the fall in the rate of profit, which we have simply listed here as examples, correspond with those mentioned in Volume III of *Capital*. The main point, however, is that Marx originally regarded the study of these moments as standing outside the analysis of 'capital in general'. Consequently, we read in the manuscript of *Theories of Surplus-Value*, which is of later origin : 'The process of the falling rate of profit would soon bring capitalist production to the point of crisis if it were not for the fact that alongside the centripetal forces, counteracting tendencies exist, which continuously exert a decentralising influence; this need not be described here, for it belongs to the chapter dealing with the competition of capitals.'[22] Marx did not devote a separate chapter to these retarding factors until Volume III of *Capital*, when it was inserted in connection with the change in the plan of the work. (Chapter 14, 'Countervailing Influences'). Despite this, even such an important moment as the devaluation of capital through crises is not dealt with here, as Marx stressed repeatedly in *Capital*[23] and in the *Theories*,[24] that 'a further analysis of crises falls outside the scope of our study'.

We have seen that in opposition to Ricardo, who attributed the tendency of the rate of profit to fall to nature,[25] Marx maintained that the fall could only be explained by the fact that 'although the worker is exploited more than, or just as much as, before . . . the portion of capital which is exchanged for living labour declines relatively'.[26] However, within certain limits capital is able to compensate for the fall in the rate of profit by increasing the mass of profit. We read on this subject in the *Rough Draft* : 'The *gross profit*, i.e. the surplus-value, regarded apart from its formal relation, not as a proportion but rather as a simple magnitude of value without connection with any other, will grow on the average *not as does the rate of profit, but as does the size of the capital*. Thus, while the rate of profit will be inversely related to the value of the capital, the *sum of profit* will be directly related to it. However, even this statement is true only for a restricted stage of the development of the productive power of capital or of labour. A capital of 100 with a profit of 10 per cent yields

21 *Grundrisse*, pp.750-51.
22 *Theories* III, p.311.
23 *Capital* III, pp.362, 831.
24 *Theories* II, pp.468, 484.
25 *Capital* III, p.242.
26 *Theories* III, p.241.

a smaller sum of profit than a capital of 1,000 with a profit of 2 per cent. In the first case the sum is 10, in the second 20, i.e. the gross profit of the larger capital is twice as large as that of the 10 times smaller capital, although the rate of the smaller capital's profit is 5 times greater than that of the larger. But if the larger capital's profit were only 1 per cent then the sum of profit would be 10, like that for the 10 times smaller capital, because the rate of profit would have declined in the same relation as its size. If the rate of profit of the capital of 1,000 were only $\frac{1}{2}$ per cent, then the sum of its profit would be only half as large as that of the smaller capital, only 5, because the rate of profit would be 20 times smaller.[27] Thus, expressed in general terms : if the rate of profit declines for the larger capital, but not in relation with its size, then the gross profit rises although the rate of profit declines. If the profit rate declines relative to its size, then the gross profit remains the same as that of the smaller capital; remains stationary. If the profit rate declines more than its size increases, then the gross profit of the larger capital decreases relative to the smaller one in proportion as its rate of profit declines.'[28]

Marx concludes by saying that the law of the tendency of the rate of profit to fall is 'in every respect the most important law of modern political economy . . . despite its simplicity, it has never before been grasped and, even less, consciously articulated . . . It is from the historical standpoint the most important law.' [29] It implies 'that the material productive power already present, already worked out, existing in the form of fixed capital, together with the scientific power, population etc., in short all conditions . . . for the reproduction of wealth, i.e. the abundant development of the social individual – that the development of the productive forces brought about by the historical development of capital itself, when it reaches a certain point, suspends the self-valorisation of capital, instead of positing it.[30] Beyond a certain point, the development of the powers of production becomes a barrier for capital; hence the capital-relation a barrier for

27 This is basically a repetition of Ricardo's arguments which Marx cited later in the *Grundrisse* and in *Capital* III, p.224. (Ricardo, *op. cit.* 142-43.)

28 *Grundrisse*, p.748.

29 Cf. *Capital* III, p.213: 'The mystery whose solution has been the goal of all political economy since Adam Smith' and *Selected Correspondence*, p.194, 'the *pons asinorum* of all previous economics'.

30 It says in the *Rough Draft* that since the falling of the rate of profit 'signifies the same as the decrease of immediate labour relative to the size of the objectified labour which it reproduces and newly posits, capital will attempt every means of checking the smallness of the relation of living labour to the size of the capital generally, hence also of the surplus-value, if expressed as profit, relative to the presupposed capital, by reducing the allotment made

the development of the productive powers of labour. When it has reached this point, capital, i.e. wage-labour, enters into the same relation towards the development of social wealth and of the forces of production, as did the guild system, serfdom, slavery, and is necessarily stripped off as a fetter. The last form of servitude assumed by human activity, that of wage-labour on one side, capital on the other, is thereby cast off like a skin, and this casting-off itself is the result of the mode of production corresponding to capital; the material and mental conditions of the negation of wage-labour and of capital, themselves already the negation of earlier forms of unfree social production, are themselves results of its production process.

'The growing incompatibility between the productive development of society and its hitherto existing relations of production expresses itself in bitter contradictions, crises, spasms. The violent destruction of capital not by relations external to it, but rather as a condition of its self-preservation, is the most striking form in which advice is given it to be gone, and to give room to a higher state of social production.'[31]

The third section of the *Rough Draft* ends with this prognosis of 'breakdown'.[32]

---

to necessary labour and by still more expanding the quantity of surplus labour with regard to the whole labour employed. Hence the highest development of productive power together with the greatest expansion of existing wealth will coincide with depreciation of capital, degradation of the labourer and a most straitened exhaustion of his vital powers.' (*Grundrisse*, p.750.)

[31] *Grundrisse*, pp.749-50. An alternative version, originally in English on p.636 of the German edition, reads as follows: 'These contradictions lead to explosions, cataclysms, crises, in which, by momentaneous suspension of labour and annihilation of a great portion of capital the latter is violently reduced to the point where it can go on . . . Yet, these regularly recurring catastrophes lead to their repetition on a higher scale, and finally to its violent overthrow.'

[32] The assertion that Marx did not propose a 'breakdown theory' is primarily attributable to the revisionist interpretation of Marx before and after the First World War. Rosa Luxemburg and Henryk Grossmann both rendered inestimable theoretical services by insisting, as against the revisionists, on the breakdown theory.

27.
Fragments on Interest and Credit

1. The extent to which the original outline envisaged the treatment of these themes

We have yet to comment on a number of pages in the *Rough Draft*, where Marx dealt with interest and interest-bearing capital.

The fact that this study is very short (in all no more than four sides – disregarding the numerous remarks which can be found throughout the manuscript) is not only explained by the haste with which Marx worked on the completion of the draft, and by the fact that he was taken ill as a result of overwork as he neared its completion,[1] but also, and primarily, by the structure of the work itself. As we know, the *Rough Draft* was not intended to go beyond the framework of 'capital in general'; from the outset this excluded a more detailed study of interest-bearing capital – not to mention the role it plays in the modern credit system. Consequently, Marx could only touch upon the category of interest in the *Rough Draft* (in connection with the study of profit and the general rate of profit), whilst, according to his original plan, the analysis of the credit system was to be held back until after the treatment of competition, i.e. in Section 3 of the *Book on Capital*.[2]

Marx also kept to this intention in his second large manuscript, that of 1862-63. Thus, we read in the section of Part III of the *Theories* which deals with profit and interest : 'This is not the place for a more detailed analysis of interest and its relation to profit; nor is it the place for an examination of the ratio in which profit is divided into industrial profits and interest.'[3] And, seven pages later : 'A general *rate of interest* corresponds naturally to the general *rate of profit*. It is not our intention to discuss this further here, since the

---

[1] Cf. Marx's letter to Engels of 29 March 1858: 'I have been very sickly again for the last two weeks with my liver. Continually working at night and a lot of petty troubles during the day, resulting from the economical conditions of my domesticity, have caused me to suffer frequent relapses.' (*MEW* Vol.29, p.309.)

[2] See Marx's outlines on pp.275 and 264 of the *Rough Draft*.

[3] *Theories* III, p.455.

analysis of interest-bearing capital does not belong to this general section[4] but to that dealing with *credit*.'[5] Accordingly, we find no analysis of credit, its role and its forms in the *Theories*, apart from the occasional comment. Marx restricts himself to demonstrating that (1) the category of interest in modern society presupposes the full development of industrial capital, and (2) that the 'alienation' of the capital-relation, its fetishisation, reaches its culmination precisely in interest-bearing capital. Apart from this, it is repeatedly stressed in the *Theories* that the analysis of credit as such can only be given at a later stage. Marx states in Part II : 'Here we need only consider the forms which capital passes through in the various stages of its development. The real conditions within which the actual process of production takes place are therefore not analysed . . . We do not examine the competition of capitals, or the credit system.'[6] (Cf. the similar passage of the same Part, according to which the 'real crisis' can only be presented from the 'real movement of capitalist production, competition and credit'.[7]) We also read, in the same Part : '*Credit* is therefore the means by which the capital of the whole capitalist class is placed at the disposal of each sphere of production, not in proportion to the capital belonging to the capitalists in a given sphere but in proportion to their production requirements – whereas in competition the individual capitals appear to be independent of each other. Credit is both the result and the condition of capitalist production, and this provides us with a convenient transition from the *competition between capitals* to *capital as credit*.'[8] This sentence is crucial for understanding Marx's plan for the construction of his whole work.

We can see that the *Theories* still keep to the original plan, which is adhered to up until Volume III of *Capital*. This latter Volume goes far beyond it and is the first occasion on which Marx steps beyond the limits of 'capital in general', in the sense in which this term was employed originally.[9] Although the first four chapters of Part V of *Capital*, Volume III do no more than develop the ideas

---

[4] That is, the section dealing with 'capital in general'.

[5] *Theories* III, p.462. In fact there is no analysis in this part of the *Theories* of the way in which the division of total profit into interest and industrial profit takes place, nor of how the relation of the rate of interest to the general rate of profit is established. Such an analysis is not to be found until Chapter XX of Volume III of *Capital*, p.358.

[6] *Theories* II, pp.492-93.

[7] *ibid.* p.512.

[8] *ibid.* p.211.

[9] Cf. pp.20ff above.

which Marx dealt with in the concluding part of the *Theories*,[10] the remaining chapters (XXV-XXXV) contain a detailed analysis of the credit system, in the short Chapter XXVII, 'in relation to industrial capital' itself, and in the further chapters 'in relation to interest-bearing capital as such'.[11] However, Engels found it necessary to redraft[12] this part of Marx's manuscript, and it is therefore difficult to say how much of it if any Marx himself had intended to use for an 'eventual continuation of the work'. Nevertheless, Chapter XXV, which deals with 'Credit and Fictitious Capital', begins with the remark : 'An exhaustive analysis of the credit system, and of the instrument which it creates for its own use (credit-money, etc.) lies outside our plan. We merely wish to dwell here upon a few particular points which are required to characterise the capitalist mode of production in general.'[13] This is stated even more categorically in Part I of *Capital* Volume III (section entitled, 'Appreciation, Depreciation, Release and Tie-Up of Capital') : 'The phenomena analysed in this chapter require for their full development the credit system and competition on the world market . . . These more concrete forms of capitalist production can only be comprehensively presented, however, after the general nature of capital is understood. Furthermore, they do not

[10] See *Theories* III, pp.499ff.

[11] 'So far we have considered the development of the credit system – and the implicit latent abolition of capitalist property – mainly with reference to industrial capital. In the following chapters we shall consider credit with reference to interest-bearing capital as such, and to its effects on this capital, and the form it thereby assumes.' (*Capital* III, pp.440-41.)

[12] Engels comments on this in the Foreword to Volume III. 'The greatest difficulty was presented by Part V, which dealt with the most complicated subject in the whole volume . . . Here there . . . was no finished draft, not even a scheme whose outlines might have been filled out, but only the beginning of an elaboration – often just a disorderly mass of notes, comments and extracts.' (*ibid.* p.4.) We further find out that only Chapters XXI-XXIX were 'in the main complete', whereas Chapters XXX to XXXIV had to be fundamentally redrafted.

[13] *ibid.* p.400. Cf. also the beginning of Chapter XXII ('Division of Profit, Rate of Interest. Natural Rate of Interest') : 'The subject of this chapter, like all the other phenomena of credit we shall come across later on, cannot be analysed here in detail. The competition between lenders and borrowers and the resultant minor fluctuations of the money market fall outside the scope of our inquiry. The circuit described by the rate of interest during the industrial cycle requires for its presentation the analysis of the cycle itself, but this likewise cannot be given here. The same applies to the greater or lesser approximate equalisation of the rate of interest in the world market. We are here concerned with the independent form of interest-bearing capital and the individualisation of interest, as distinct from profit.' (*ibid.* p.358.)

come within the scope of this work and belong to its eventual continuation.'[14]

2. The 'Rough Draft' on interest-bearing capital

Be that as it may, what has been stated is sufficient to explain the fragmentary character of the observations which Marx makes on interest-bearing capital and the credit system in the *Rough Draft*. All that he is concerned to do here is firstly to show that the development of capital itself must lead to the division of surplus-value into industrial profit and interest, and to the 'autonomisation of interest in relation to profit'. And secondly, that the analysis of 'capital in general' already contains the seeds of the basic definitions, from which the theory of credit can be developed.

However, doesn't the category of interest considerably pre-date that of profit? And doesn't Marx always stress the 'historical pre-existence of interest-bearing capital' in comparison with industrial capital proper?[15]

We know that it is in its particular function as the means of payment that money 'develops interest and hence money-capital'.[16] Simple commodity circulation already produces relations 'under which the alienation of the commodity becomes separated by an interval of time from the realisation of its price ... This gives rise to relations of creditor and debtor among commodity owners. These relations can be fully developed even before the credit system comes into being, although they are the natural [and spontaneously arisen] basis of the latter.'[17] For 'there was borrowing and lending in earlier situations as well, and usury is even the oldest of the antediluvian forms of capital. But borrowing and lending no more constitute *credit* than working constitutes *industrial labour or free wage-labour*. And credit as an essential, developed relation of production appears *his-*

---

[14] *Capital* III, Part I, Chapter VI, Section II, p.110.

[15] *ibid.* p.367. Cf. *ibid.* p.376: 'Yet historically interest-bearing capital existed as a completed, traditional form, and hence interest as a completed subdivision of the surplus-value produced by capital, long before the existence of the capitalist mode of production and its attendant conceptions of capital and profit.'

[16] *ibid.* p.598. Money dealing can be characterised as the second origin of the credit system, in connection with which 'the management of interest-bearing capital ... develops as a special function of the money-dealers.' (*ibid.* p.402.)

[17] *Capital* I, pp.232-33 (134-35) and *Contribution*, p.143.

torically only in circulation based on capital or on wage-labour . . .
Although *usury* is itself a form of credit in its *bourgeoisified* form,
the form *adapted to capital*, in its pre-bourgeois form it is rather the
expression of lack of credit.[18]

The issue is therefore that of the differing social role played by
interest-bearing capital in capitalism and in pre-capitalist situations.
'The presentation of the specific, distinguishing characteristics', states
Marx in opposition to Storch's comments on credit, 'is here both the
logical development and the key to the understanding of the *historical* development'.[19] What 'distinguish interest-bearing capital
from usurer's capital, insofar as it is an essential element of the capitalist mode of production', are primarily 'the altered conditions under
which it operates, and consequently also the totally transformed
character of the borrower who confronts the money-lender'. The
usurer lends, firstly, to small producers, who own the conditions of
their labour (artisans, and above all peasants), and, secondly, to
'extravagant members of the upper classes', basically landowners; the
modern bank lends to capitalists. 'Even when a man without fortune
receives credit in his capacity of industrialist or merchant, it occurs
with the expectation that he will function as capitalist, and appropriate unpaid labour with the borrowed capital. He receives credit
in his capacity of potential capitalist.' On the other hand, modern
credit presupposes the full development of commodity production
and circulation. The opposite is the case with usury. 'The more insignificant the role played by circulation in social reproduction, the
more usury flourishes.'[20]

The above shows how absurd it is to lump together the interest-
bearing capital of today with its 'antediluvian' form. We read in
the *Rough Draft* : 'The level of interest in India for common agricul-
turalists in no way indicates the level of profit. But, rather, that profit,
as well as part of the wage itself, is appropriated in the form of
interest by the usurer.[21] It requires a sense of history like that of Mr.
Carey to compare this interest with that prevailing on the English
money market, which the English capitalist pays, and to conclude
therefrom how much higher the "labour-share" (the share of labour
in the product) is in England than in India. He ought to have com-

---

[18] *Grundrisse*, p.535.
[19] *ibid.* p.672.
[20] *Capital* III, pp.594, 600, 610.
[21] It is evident that Marx is speaking only of the 'embryonic' forms of
profits and wages here, since he is concerned with pre-capitalist conditions.

388 · *The Making of Marx's 'Capital'*

pared the interest which English handloom weavers, e.g. in Derby-
shire, pay, whose material and instrument is advanced (lent) by the
capitalist. He would have found that the interest is here so high that,
after settlement of all items, the worker ends up being the debtor,
after not only having made restitution of the capitalist's advance, but
also having added his own labour to it free of charge.'[22] Furthermore,
Carey should have seen that 'historically . . . the form of industrial
profit arises only after capital no longer appears alongside the independent worker. Profit thus appears originally' i.e. in pre-capitalist
situations, 'as determined by interest. But in the bourgeois economy,
interest [is] determined by profit, and [is] only one of the latter's
parts. Hence profit must be large enough to allow a part of it to branch
off as interest. Historically, the inverse. Interest must have become
so depressed that a part of the surplus gain could achieve independence as profit.' And further : 'Where this relation' of the independent,
small-scale producer, who is, however, afflicted by usury, 'repeats
itself within the bourgeois economy, it does so in the backward
branches of industry, or in such branches as still struggle against their
extinction and absorption into the modern mode of production. The
most odious exploitation of labour still takes place in them, although
here the relation of capital and labour does not carry within itself
any basis whatever for the development of new forces of production,
or the germ of newer historic forms. In the mode of production itself,
capital still here appears materially subsumed under the individual
workers or the family of workers – whether in a handicraft business
or in small-scale agriculture. What takes place is exploitation by
capital without the mode of production of capital . . . This form of
usury, in which capital does not seize possession of production, hence
is capital only formally, presupposes the predominance of pre-
bourgeois forms of production; but reproduces itself again in subordinate spheres within the bourgeois economy itself.'[23]

What must therefore be stressed at the outset is that 'interest and
profit both express relations of *capital*'; that the category of interest
'presupposes the division of *profit* into interest and profit'. 'The difference,' says Marx, 'becomes tangible, perceptible as soon as a class
of monied capitalists comes to confront a class of industrial capital-

---

[22] Cf. *Capital* III, p.597 : 'For instance, if we wish to compare the
English interest rate with the Indian, we should not take the interest rate of
the Bank of England, but rather, e.g. that charged by lenders of small
machinery to small producers in domestic industry.'

[23] *Grundrisse*, pp.851-53. This passage can be found, redrafted and
expanded, in *Capital* III, pp.595-98.

ists'.[24] However, monied capitalists and industrial capitalists can only 'form two particular classes because profit is capable of separating off into two branches of revenue'. The mere existence of these classes 'presupposes a division within the surplus-value posited by capital'.[25]

3. The category of 'capital as money'

The possibility of this internal division of surplus-value results from the very fact of the valorisation of capital. That is, after the money advanced by the capitalist in the production process has been valorised, it takes on 'the *new* aspect of realised capital'; it turns into the 'perpetually valid form of appearance of capital'.[26] It is of course true that it 'exists, objectively, merely as *money*'; but this money 'is *in itself* already capital; and, as such, it is a *claim on new labour*. Here capital already no longer enters into relation with ongoing labour, but with future labour . . . As a claim, its material existence as money is irrelevant, and can be replaced by any other title. Like the creditor of the state, every capitalist with his newly gained value possesses a claim on future' (alien) 'labour, and, by means of the appropriation of ongoing labour has already at the same time appropriated future labour.' (Marx adds : 'This side of capital to be developed to this point. But already here its property of existing as value separately from its substance can be seen. This already lays the basis for credit.') Thus for the capitalist 'to stockpile it in the form of money is . . . by no means the same as materially to stockpile the material conditions of labour.This is rather a stockpiling of property titles to labour. Posits future labour as *wage-labour*, as use-value for capital.'[27] Only in this way is it possible for 'capital itself . . . to become a commodity', or for 'the commodity (money) to be sold as capital'.[28]

We thus come to the category of 'capital as commodity' or 'capital as money', as distinct from the category of 'money as capital'

---

[24] In this sense we read in *Capital*: 'It is indeed only the separation of capitalists into money-capitalists and industrial capitalists that transforms a portion of the profit into interest, that creates the category of interest at all.' (*Capital* III, p.370.)

[25] *Grundrisse*, pp.852-53.

[26] *ibid.* p.447.

[27] *ibid.* p.367.

[28] *ibid.* p.851.

expounded previously.[29] In Section I of the *Rough Draft* we read the following : 'As interest, capital itself appears again in the character of a *commodity*, but a commodity *specifically* distinct from all other commodities; *capital as such* – not as a mere sum of exchange-values – enters into circulation and becomes a *commodity*. Here the character of the commodity (i.e. the particular use-value of capital) is itself present as an *economic, specific* determinant, not irrelevant as in simple circulation, nor directly related to labour as its opposite, as its [i.e. capital's] use-value, as with industrial capital[30] . . . The commodity as capital, or capital as *commodity*, is therefore not exchanged for an equivalent in circulation; by entering into circulation, it obtains *its being-for-itself*;[31] it obtains its original relation to its owner, even when it passes into the possession of another. It is therefore merely *loaned*. For its owner, its use-value as such is its valorisation; money as money, not as medium of circulation; its *use-value* as *capital.*'[32] Or, as we read in the *Theories* : 'Since, on the basis of capitalist production, a certain sum of values . . . makes it possible to extract a certain amount of labour gratis from the workers and to appropriate a certain amount of surplus-value, surplus labour, surplus-product, it is obvious that money itself can be sold as capital, that is, as a commodity *sui generis* . . . It can be sold as the source of profit. I enable someone else by means of money, etc. to appropriate surplus-value. Thus it is quite in order for me to receive part of this surplus-value. Just as land has value because it enables me to intercept a portion of surplus-value, and I therefore pay for this land only the surplus-value which can be intercepted thanks to it, so I pay for capital the surplus-value which is created by means of it. Since, in the capitalist production process, the value of capital is perpetuated and reproduced in addition to its surplus-value, it is therefore quite in order that, when money or commodities are sold as capital, they return to the seller after a period of time and he does not alienate

[29] Cf. p.186 above.
[30] Cf. Chapter 3, pp.80-83 above.
[31] See the note on p.244 of the *Grundrisse*.
[32] *ibid.* pp.318-19. 'What, now, is the use-value which the money-capitalist gives up for the period of the loan and relinquishes to the productive capitalist – the borrower? It is the use-value which the money acquires by being capable of becoming capital, of performing the functions of capital, and creating a definite surplus-value, the average profit . . . during its process, besides preserving its original magnitude of value. In the case of the other commodities the use-value is ultimately consumed. Their substance disappears, and with it their value. In contrast, the commodity-capital is peculiar in that its value and the use-value not only remain intact, but also increase through consumption of its use-value.' (*Capital* III, p.351.)

them in the same way as he would a commodity. In this way, money or commodities are not sold as money or commodities, but in their second power, as *capital*, as self-augmenting money or commodity-value.'[33]

4. Critique of Proudhonism

It is precisely the fact that capital, in that it becomes a commodity, can only be loaned, and must therefore return to its owner, which forms the basis of the critique of interest-bearing capital by the petit-bourgeois socialists (Proudhon and his school). Marx writes : 'In the whole polemic by M. Proudhon against Bastiat ... Proudhon's argument revolves around the fact that lending appears as something quite different to him from selling. To lend at interest' (thinks Proudhon), 'is "the ability of selling the same object again and again, and always receiving its price anew, without ever giving up owner-ship of what one sells" ... The different form in which the repro-duction of capital appears here deceives him into thinking that this constant reproduction of the capital – whose price is always obtained back again, and which is always exchanged anew for labour at a profit, a profit which is realised again and again in purchase and sale – constitutes its concept. What leads him astray is that the "object" does not change owners, as with purchase and sale; thus basically only the form of reproduction peculiar to capital lent at interest changes into the form of reproduction peculiar to fixed capital.' However : 'If the circulating capital is regarded in its whole process, then it may be seen that, although *the same object* (this specific pound of sugar e.g.) is not always sold anew, the same value does always reproduce itself anew, and the sale concerns only the form, not the substance.' Thus, according to Proudhon 'everything should be *sold*, not *lent*.' He 'wants to cling to the simplest, most abstract form of exchange', at the same time failing to grasp that 'the exchange of commodities rests on the exchange between capital and labour', and that not only does the category of profit proceed neces-sarily from this very exchange, but also that of interest. That is, he does not understand that 'in order to abolish interest, he also has to abolish *capital* itself, the mode of production based on exchange-value, and hence wage-labour too'.[34] His demand 'that capital should

---

[33] *Theories* III, p.455.
[34] *Grundrisse*, pp.843-44. (The same passage – redrafted – appears in Volume III of *Capital*, pp.345-47.)

not be loaned out and should bear no interest, but should be sold like a commodity for its equivalent, amounts at bottom to no more than the demand that exchange-value should never become capital, but always remain simple exchange-value; that *capital* should *not exist as capital*. This demand, combined with the other, that wage-labour should remain the general basis of production, reveals a happy confusion with regard to the simplest economic concepts.'[35]

5. The 'Rough Draft' on the role of credit in the capitalist economy

At this point we must be careful to make a distinction between the possibility and the necessity of the credit system.

We saw how the possibility of credit relations originated in money's function as means of payment. We saw, further, how on the basis of the capitalist mode of production any sum of money capable of being laid out as capital represented a 'claim to alien labour', and could therefore be lent at interest as a potential source of profit. The actual provision of such loan-capital, on a regular basis and in sufficient quantity, is taken care of by the circulation process of capital, within which sums of money are periodically released, which are superfluous to the requirements of the individual's own enterprise and can therefore be put at the disposal of other capitalists through the medium of credit.[36]

The possibility of credit is therefore a product of the 'inner nature' of the capitalist mode of production itself; it is inherent to its 'concept'. There are, however, moments in the development and life-cycle of capital, which establish not only the possibility, but also the necessity of the credit system; which in fact cause credit to appear as a necessary condition of capitalist production; the chief of these is the striving for continuity, for the uninterrupted flow of the production process.

Why this striving is necessary is quite obvious. Capital only creates surplus-value in the production process : 'the *constant continuity*' of this process therefore 'appears as a fundamental condition for production based on capital'. On the other hand, each phase of

---

[35] *Grundrisse*, p.319.
[36] 'The money-capital thus released by the mere mechanism of the turnover movement (together with that freed by the successive reflux of fixed capital and that required in every labour process for variable capital) must play an important role as soon as the credit system develops and must at the same time form one of the latter's foundations.' (*Capital* II, p.286.)

production must be followed by a phase of circulation, which constantly interrupts the continuity of production. 'Thus the conditions of production arising out of the nature of capital itself contradict each other. The contradiction can be superseded and overcome only in two ways'. First, by the division of capital into portions (discussed in Chapter 23 above); and secondly – through credit. 'A pseudo-buyer B – i.e. someone who really *pays* but does not really buy – mediates the transformation of capitalist A's products into money. But B himself is paid only after capitalist C has brought A's products. Whether the money which this credit-man B gives to A is used by A to buy labour or to buy raw material and instrument, before A can replace either of them from the sale of his product, does not alter the case . . . In this case capital B replaces capital A; but they are not valorised at the same time. Now B takes the place of A; i.e. his capital lies fallow, until it is exchanged with capital C. It is frozen in the product of A, who has made his product liquid in capital B.'[37]

According to Marx this is an aspect of credit which 'arises out of the direct nature of the production process', and thus constitutes 'the foundation of the necessity for credit'.[38] But the other moments which establish the necessity of credit are no less important.

We know that circulation time is always a barrier to the creation and realisation of value, 'a barrier arising not from production generally but specific to the production of capital'.[39] Consequently the 'necessary tendency of capital', is not only to shorten circulation time, but to reduce it to nothing wherever possible, i.e. to bring about '*circulation without circulation time*'. Marx stresses that it is precisely this tendency which is 'the fundamental determinant of credit and of capital's credit contrivances'.[40] In this context we should refer first to the function of money as a 'circulation machine', which is bound up with large unproductive expenditure. 'Insofar as it is value in itself' it is to be characterised as 'one of the principal costs of circulation' of capitalist production.[41] Hence capital strives to 'economise'

---

[37] *Grundrisse*, pp.534-35, 549.
[38] *ibid.* p.535.
[39] *ibid.* p.543.
[40] *ibid.* p.659.
[41] *Capital* III, p. 435. Cf. *Capital* II, p.350: 'The entire amount of labour-power and social means of production expended in the annual production of gold and silver intended as instruments of circulation constitutes a bulky item of the *faux frais* of the capitalist mode of production, of the production of commodities in general. It withdraws from social utilisation as many additional means of production and consumption, i.e. as much real wealth as possible. To the extent that the costs of this expensive machinery of circulation are decreased, the given scale of production or the given degree of its extension

on money, and to posit it 'as a merely formal moment : so that it mediates the formal transformation' of commodities, 'without itself being *capital*, i.e. value';[42] on the other hand the striving 'to give *circulation time value*, the value of *production-time*, in the various organs which mediate the process of circulation time and of circulation; to posit them all as money, and, more broadly as capital . . . All this springs from the same sources. All the requirements of circulation . . . although they take on different and seemingly quite heterogeneous forms, are all derived from *circulation time*. The machinery for abbreviating it is itself a part of it.' And it is precisely for this reason that 'the contradiction of labour-time and circulation time contains the entire doctrine of credit, to the extent, namely, that the history of currency etc. enters here.'[43]

However, circulation time is not the only barrier against which capital's drive for valorisation collides. There is also the barrier presented by the sphere of exchange; this consists in the fact that on the one hand capital must produce on a capitalist basis without regard to the limited dimensions of consumption, but that on the other, as value, it presupposes a counter-value against which it can be exchanged.[44] Credit is also enormously significant in this context – as the course of the industrial cycle shows. In fact, adds Marx, 'this appears more colossally classically, in the relation between peoples than in the relation between individuals. Thus e.g. the English forced to *lend* to foreign nations, in order to have them as customers. At bottom the English capitalist exchanges doubly with *productive* English capital, (1) as himself, (2) as Yankee etc. or in whatever other form he has placed his money.' [45]

remaining constant, the productive power of social labour is, *eo ipso*, increased. Hence, so far as the forms of existence which develop with the credit system have this effect, they increase capitalist wealth directly, either by performing a large portion of the social production and labour process without any intervention of real money, or by improving the ability of the quantity of money actually functioning to perform its functions.'

[42] Marx stresses in *Capital* that 'it should always be borne in mind that money – in the form of precious metal – remains the foundation, from which the credit system, by its very nature, can never detach itself.' (*Capital* III, p.606.)

[43] *Grundrisse*, pp.659-60.

[44] This necessity clearly would not exist 'if all capitals [produced] to order for each other, and the product [was] therefore always immediately money' : but this is a 'notion which contradicts the nature of capital, and hence also the practice of large-scale industry'. (*ibid.* p.549.)

[45] *ibid.* p.416. Cf. *Theories* III, p.122 : 'The author [of *An Inquiry into Those Principles* . . .] also admits that the credit system may be a cause of crises (as if the credit system itself did not arise out of the difficulty of employ-

(In *Capital* III Marx refers in addition to the 'necessary development' of credit, 'in order to effect the equalisation of the rate of profit . . . upon which the whole of capitalist production rests';[46] however, this aspect is not mentioned in the *Rough Draft*.)

6. The barriers of the credit system

As we have just seen, 'the entire *credit system*, and the over-trading, over-speculation etc. connected with it, rests on the necessity of expanding and leaping over the barrier to circulation and the sphere of exchange'.[47] It is in this sense that credit is 'an inherent form of the capitalist mode of production' upon which 'the entire connection of the reproduction process is based'.[48] Nevertheless, the role of credit should not be overestimated! For, just as 'money overcomes the barriers of barter only by generalising them – i.e. separating sale and purchase entirely', credit, similarly, only overcomes the barriers to the valorisation of capital 'by raising them to their most general form, positing one period of overproduction and one of underproduction as two periods'.[49] Although its development results in 'the acceleration . . . of the individual phases of circulation, or of the metamorphosis of commodities, and with it an acceleration of the process of reproduction in general', at the same time 'credit helps to keep the acts of buying and selling further apart in time and serves thereby as the basis for speculation'.[50] For this reason Marx mocks

ing capital "productively", i.e. "profitably"). The English, for example, are forced to lend their capital to other countries in order to create a market for their commodities. Overproduction, the credit system, etc. are means by which capitalist production seeks to break through its own barriers and to produce over and above its own limits. Capitalist production, on the one hand, has this driving force; on the other hand, it only tolerates production commensurate with the profitable employment of existing capital. Hence crises.'

[46] *Capital* III, p.435.
[47] *Grundrisse*, p.416.
[48] *Capital* III, p.606.
[49] *Grundrisse*, p.623.
[50] *Capital* III, p.436. Cf. *ibid.* p.441: 'The credit system appears as the main lever of overproduction and overspeculation in commerce solely because the reproduction process, which is elastic by nature, is here forced to its extreme limits, and is so forced because a large part of the social capital is employed by people who do not own it and who consequently tackle things quite differently than the owner, who anxiously weighs the limitations of his private capital insofar as he handles it himself. This simply demonstrates the fact that the valorisation of capital based on the contradictory nature of capitalist production permits real, free development only up to a certain

the 'circulation artists', 'who imagine that they can do something with the velocity of circulation other than lessen the obstacles to reproduction . . . Even madder, of course, are those circulation artists who imagine that credit institutes and inventions which abolish the lag of circulation time will not only do away with the delays and interruptions in production caused by the transformation of the finished product into capital, but will also make the capital, with which productive capital exchanges, itself superfluous; i.e. they want to produce on the basis of exchange-value but at the same time to remove, by some witchcraft, the necessary conditions of production on this basis. The most that credit can do in this respect – as regards *mere* circulation – is maintain the continuity of the production process, *if* all other conditions of this continuity are present, i.e. if the capital to be exchanged with actually exists etc.'[51]

These arguments from the *Rough Draft* are clearly still valid today. The same also applies to Marx's critique of the 'illusions concerning the miraculous power of the credit and banking system in the socialist sense': 'As soon as the means of production have ceased to be transformed into capital (which also includes the abolition of private property in land), credit as such no longer has any meaning . . . On the other hand, as long as the capitalist mode of production continues to exist, interest-bearing capital, as one of its forms, also continues to exist and constitutes in fact the basis of its credit system.'[52] And consequently, the conception of the essentially 'socialist' character of credit belongs in the arsenal of petit-bourgeois utopias. This is clearly not contradicted by the fact that credit reveals itself as the driving force in the development of the capitalist order to its 'highest and ultimate form'[53] and hence works towards its dissolution; the reason is that it is credit which represents the form 'in which capital tries to posit itself as distinct from the individual capital' and in which the social character of capitalist production is most strikingly

point, so that in fact it constitutes an immanent fetter and barrier to production, which is continually broken through by the credit system. Hence, the credit system accelerates the material development of the productive forces and the establishment of the world market. It is the historical mission of the capitalist system of production to raise these material foundations of the new mode of production to a certain degree of perfection. At the same time credit accelerates the violent eruptions of this contradiction – crises – and thereby the elements of disintegration of the old mode of production.'

[51] *Grundrisse*, pp.545-46.
[52] *Capital* III, p.607.
[53] *ibid.*

expressed.[54] 'The highest result it [i.e. capital] achieves in this line is, on one side, *fictitious capital*; on the other side, credit only appears as a new element of *concentration*, of the destruction of capitals by individual, centralising capitals.'[55] However, this raises a series of questions which go far beyond the study of 'capital in general', and which are therefore not dealt with in detail in the *Rough Draft*. We should bear in mind that the reason for this exclusion is that the basic tendencies of credit are only dealt with in the *Rough Draft* as they emerge from the abstract, general analysis of the capitalist process of production and circulation.[56] It is nonetheless amazing to see the extent to which many of the later results of the analysis of the credit system in Volume III of *Capital* were anticipated by this method in the *Rough Draft*.

[54] *Grundrisse*, p.659. See Marx's outline in a letter to Engels of 2 April 1858: 'c) Credit, here capital as the general principle confronts the individual capitals.' (*Selected Correspondence*, p.97.)

[55] *Grundrisse*, p.659. Cf. *ibid.* p.657: 'This suspension (of the seeming independence and independent survival of individual capitals) takes place even more in credit. And the most extreme form to which the suspension proceeds, which is however at the same time the ultimate positing of capital in the form adequate to it – is joint-stock capital.' We have already pointed out in Chapter 2 above that Marx was able to predict the transition from competitive capitalism to monopoly capitalism in the *Rough Draft*.

[56] This explains why such an important aspect as the role of credit in the equalisation of the general rate of profit was not dealt with in the *Rough Draft*.

Appendix

On recent criticisms of Marx's law of the falling rate of profit

I. There can scarcely be one basic principle in Marx's entire system which has been so unanimously rejected by both academic and non-academic critics as his law of the tendency of the rate of profit to fall : yet the arguments used by his critics on the issue are perhaps the least satisfactory of any. Two writers who have recently concerned themselves with Marx's law can be cited as examples of this. They are Joan Robinson[1] and the American marxist, Paul Sweezy.[2]

As with earlier critics of the law, Robinson and Sweezy consider that Marx is guilty of what is primarily a methodological inconsistency. Marx allegedly formulated his law under the assumption of a constant rate of surplus-value; he arbitrarily separated the factors which reduce the rate of profit from those which increase it, in order to derive his law from the former, and the 'countervailing influences' from the latter. Or, as the original proponent of this objection, von Bortkiewicz, stated : 'The error in the proof of the law which Marx offers chiefly consists in the fact that he neglected the mathematical relation between the productivity of labour and the rate of surplus-value. He regarded the latter as a separate factor. The absurdities to which such a way of isolating factors generally leads can be seen from the following simple example. Take a positive value a, which is related to other positive values b and c through the function $a = b/c$.

The question is, how does a change, when each of the values b and c depend on d. For example where $b = d^5$, and $c = d^3$. The correct solution to the question is clearly this. One eliminates b and c from the expression for a, finds that $a = d^2$ and concludes from this that a varies in the same direction as d. But if Marx's method of isolating is used in this example, one could express e.g. a by b/d^3 and draw the conclusion from this formula, that a gets smaller as d increases, and larger as d decreases. If one were to add to this that a

---

[1] J.Robinson, *An Essay on Marxian Economics*, London: Macmillan 1966, Chapter 5.
[2] P.M.Sweezy, *The Theory of Capitalist Development*, 1942, Chapter 6.

change in *b* could obscure this connection, but that this is a separate thing, then the essential identity of this procedure with Marx's method of isolating becomes even clearer.'[3]

Is this objection valid? Did Marx really lay himself open to the accusation that he violated the elementary rules of logic? Let us take a closer look.

II. The first (but only the first) page of Chapter XIII in Volume III of *Capital*, which deals with the law of the tendency of the rate of profit to fall, seems, in fact, to prove the correctness of the above statement. Marx begins with a numerical example intended to show how differences in the organic composition of capital affect the particular rates of profit in the five different branches of production (before the equalisation of these rates of profit to an average rate). Naturally, this can be shown most easily if one temporarily abstracts from other factors which could influence the rate of profit – chiefly differences in the degree of exploitation of labour. Consequently, Marx assumes – as in the previous sections in Volume III – that the rate of surplus-value is equal to 100 per cent in all five branches of production; i.e. that the workers work half a day for themselves and half a day for the employer. Marx demonstrates that the rates of profit in the five branches of production have to stand in an inverse relation to the level of the organic composition.[4] However, what

---

[3] L. von Bortkiewicz, *'Wertrechnung und Preisrechnung im Marxschen System'* in *Archiv für Sozialwissenschaft und Sozialpolitik*, Sept. 1907, pp.466-67. (English translation 'Value and price in the marxian system', *International Economic Papers* no.2, 1952.) In addition Bortkiewicz should also have directed his criticism not only at Marx, but also at Mill, since Mill had already dealt with the problem of the tendency of the rate of profit to fall in two stages; first he looked at the law itself; and then, the factors counteracting it. (It was Henryk Grossmann who first referred to the similarity in the methods used by Marx and Mill in his *Akkumulations-und Zusammenbruchsgesetz des kapitalistischen Systems*, p.116.)

[4] Marx gives the following example (*Capital* III, p.211):

	Constant Capital	Variable Capital	Surplus -Value	Rate of Surplus-Value	Rate of Profit
I	50	100	100	100%	66⅔%
II	100	100	100	100%	50%
III	200	100	100	100%	33⅓%
IV	300	100	100	100%	25%
V	400	100	100	100%	20%

However it is immediately clear that Marx could have constructed his example so that the rate of surplus-value increased from one branch of production to the next, e.g.

O

applies to branches of production standing alongside one another at one point in time also applies to the various successive states of the total social capital. The average composition of social capital continually increases, and this is precisely the reason why 'the gradual growth of constant capital in relation to variable must necessarily lead to a *gradual fall of the general rate of profit*, so long as the rate of surplus-value, or the intensity of exploitation of labour by capital, remains the same.'[5] But we also read on the very same page : 'The hypothetical series [of five branches of production] . . . expresses the actual tendency of capitalist production. This mode of production produces a progressive relative decrease of the variable capital as compared to the constant capital, and consequently a continually rising organic composition of the total capital. The immediate result of this is that the rate of surplus-value, at the same, or even a rising, degree of labour exploitation, is represented by a continually falling rate of profit.'[6] And two pages further on : 'The law of the falling rate of profit, which expresses the same, or even a higher rate of surplus-value, states, in other words, that any quantity of the average social capital, say, a capital of 100, comprises an ever larger portion of the means of labour, and an ever smaller portion of living labour. Therefore, since the aggregate mass of of living labour operating the means of production decreases in relation to the value of these means of production, it follows that the unpaid labour and the portion of value in which it is expressed must decline as compared to the value of the advanced total capital. Or : an ever smaller aliquot part of total invested capital is converted into living labour, and this total

	Constant Capital	Variable Capital	Surplus -Value	Rate of Surplus-Value	Rate of Profit
I	50	100	100	100%	66⅔%
II	100	100	130	130%	65%
III	200	100	192	192%	64%
IV	300	100	252	252%	63%
V	400	100	310	310%	62%

Thus, in this case too, the rate of profit would gradually fall – despite the sharply rising rate of surplus-value. (Admittedly, the example is quite arbitrary; if we had assumed a faster growth in the rate of surplus-value, then the rate of profit would not have fallen, but rather risen. However, it would be completely wrong to think that the fall in the rate of profit can be compensated for, in all circumstances, by the increase in the rate of surplus-value. We will see later why Marx rejected such an assumption from the outset.)

[5] *Capital* III, p.212.
[6] *ibid.* pp.212-13.

capital, therefore, absorbs in proportion to its magnitude less and less surplus labour, although the unpaid part of the labour applied may at the same time grow in relation to the paid part.'[7]

Marx says the same thing on pages 220-21, 226-27, 229, 233-34 and 241-42 of Volume III. And finally at the end of Chapter XIV he regarded it as necessary – 'in order to avoid misunderstandings' – to repeat : 'The tendency of the rate of profit to fall is bound up with a tendency of the rate of surplus-value to rise, hence with a tendency for the rate of labour exploitation to rise . . . The rate of profit does not fall because labour becomes less productive, but because it becomes more productive. Both the rise in the rate of surplus-value and the fall in the rate of profit are but specific forms through which growing productivity is expressed under capitalism.'[8]

The passages above are also complemented by equally categorical statements in the *Theories*.[9] Marx clearly had no intention of restricting his law to the case of a constant rate of surplus-value. According to him, even a rising rate of surplus-value must eventually be expressed in a falling rate of profit. However, this does not prevent his critics from interpreting the law in a completely different sense. Thus, Joan Robinson writes on this law : 'Marx's theory, as we have seen, rests on the assumption of a constant rate of exploitation.' This

---

[7] *ibid.* pp.215-16.

[8] *ibid.* p.240. Cf. *Theories* II, p.439: 'The rate of profit falls although the rate of surplus-value remains the same or rises, because the proportion of variable capital to constant capital decreases with the development of the productive power of labour. The rate of profit thus falls, not because labour becomes less productive, but because it becomes more productive. Not because the worker is less exploited, but because he is more exploited, whether the absolute surplus time grows or, when the state prevents this, the relative surplus time grows, for capitalist production is inseparable from falling relative value of labour.'

[9] We should also mention in this context pp.240, 302, 311 and 369 in Part III of *Theories*. On p.302 Marx states: 'I have explained the decline in the rate of profit in spite of the fact that the rate of surplus-value remains the same or even rises, by the decrease of the variable capital in relation to the constant, that is, of the living present labour in relation to the past labour which is employed and reproduced.' And on p.311 : 'This is where Hodgskin's view merges with the general law which I have outlined. The surplus-value, i.e. the exploitation of the worker, increases, but, at the same time, the rate of profit falls because the variable capital declines as against the constant capital, because, in general, the amount of living labour falls relatively in comparison with the amount of capital which sets it in motion. A larger portion of the annual product of labour is appropriated by the capitalist under the signboard of capital, and a smaller portion under the signboard of profit.'

then leads on to the following objection :[10] 'This proposition [i.e. Marx's law] stands out in startling contradiction to the rest of Marx's argument. For if the rate of exploitation tends to be constant, real wages tend to rise as productivity increases. Labour receives a constant proportion of an increasing total. Marx can only demonstrate a falling tendency in profits by abandoning his argument that real wages tend to be constant. This drastic inconsistency he seems to have overlooked.'

But, secondly, Robinson believes that Marx's law consists 'simply in the tautology : when the rate of exploitation is constant, the rate of profits falls as capital per man increases. Assuming constant periods of turnover, so that c + v measures the stock of capital : when s/v is constant and c/v is rising, s/(c+v) is falling.'[11] It is hardly surprising that Joan Robinson comes to the totally negative judgement that Marx's 'explanation of the falling tendency of profits explains nothing at all'.[12]

III. It is clear that we can now simply brush both objections aside; since Marx in no way connected his law to the assumption of a constant rate of surplus-value, he cannot be criticised either for an 'inconsistency' or a 'tautology' on this question. However, this does not completely take care of the criticisms directed against his 'method of isolating factors'. For, if Marx's law is in fact based on the assumption of a constant rate of surplus-value, why does he treat the factors which raise the rate of profit as 'things for themselves'? Why does he

---

[10] Sweezy argues in similar fashion : 'We have seen that the tendency of the rate of profit to fall is deduced by Marx on the assumption that the organic composition of capital rises while the rate of surplus-value remains constant. Is it justifiable, however, to assume at the same time a constant rate of surplus-value? It is necessary to be clear about the implications of the latter assumption. A rising organic composition of capital goes hand in hand with increasing labour productivity. If the rate of surplus-value remains constant, this means that a rise in real wages takes place which is exactly proportional to the increase in labour productivity. Suppose that labour productivity is doubled, that is to say, that in the same time labour produces twice as much as previously. Then, since an unchanged rate of surplus-value means that the labourer works the same amount of time for himself and the same amount for the capitalist as previously, it follows that both the physical output represented by the wage and the physical output represented by the surplus-value have also doubled. In other words, the labourer benefits equally with the capitalist in the increased productivity of his labour. While there can be no logical objection to an assumption which leads to this result, there are nevertheless grounds for doubting its appropriateness.' (*Theory of Capitalist Development*, pp.100-01.)

[11] J.Robinson, *Essay on Marxian Economics*, p.36.

[12] *ibid.* p.42.

leave the study of such important factors as the 'increase in the degree
of labour exploitation', or 'relative surplus population' etc. until after
the presentation of the law itself : i.e. until Chapter XIV, and merely
allot them the role of 'countervailing influences' to the law?

This objection plays a large part in Sweezy's criticism : 'It seems
hardly wise to treat an integral part of the process of rising productivity' (Sweezy means the increase in the rate of surplus-value) 'separately and as an off-setting factor; a better procedure is to recognise
from the outset that rising productivity tends to bring with it a higher
rate of surplus-value. Furthermore, this is what Marx usually does.'[13]

This is really a strange argument. The problem which confronted Marx was this : how does the continually growing productivity of social labour affect the average rate of profit? Since the level
of the rate of profit depends on two factors – the rate of surplus-
value and the organic composition of capital – and since the growth
in the productivity of labour affects both factors, then we would
indeed be presented with the 'method of isolation', criticised by
Bortkiewicz, if Marx only looked at the increase in organic composition, for example, without noting that the growing productivity of
labour must simultaneously raise the rate of surplus-value (even if to
a lesser extent), or, conversely, if he saw only the increase in the rate
of surplus-value resulting from the growth of productivity, and left
out of consideration the even stronger tendency towards an increase
in the organic composition of capital which is bound up with it. There
are also, of course, moments which only affect one of the two factors,
without immediately affecting the other. For example, any attentive
reader of the chapter dealing with 'countervailing influences' (Chapter XIV) will notice that Marx only looks at those methods of
exploitation (for example Section I – 'Increasing intensity of exploitation') in which a rise in the rate of surplus-value is not simultaneously
accompanied by an 'increase or proportional increase of constant
capital as against variable'; i.e. where the organic composition
remains initially unchanged.[14] By contrast, methods which 'include
a growth in constant capital as against variable i.e. a fall in the rate

---

[13] *Theory of Capitalist Development*, p.101.

[14] *Capital* III, p.234. 'There are many ways of intensifying labour which
imply an increase of constant, as compared to variable capital, such as when
a worker has to operate a larger number of machines . . . But there are other
aspects of intensification, such as the greater velocities of machinery, which
consume more raw material in the same time, but, as far as the fixed capital is
concerned, wear out machinery so much faster, and yet do not in any way
affect the relation of its value to the price of the labour which sets it in motion.'

of profit' (that is, what are essentially methods for producing relative surplus-value) were of course already considered in the exposition of the law itself in Chapter XIII. And for this reason 'relative over-population' is examined in Chapter XIV only insofar as it facilitates the continued existence of branches of production with an especially low organic composition, as a consequence of the 'cheapness and abundance of disposable or unemployed wage-labourers, and of the greater resistance which some branches of production, by their very nature, oppose to the transformation of manual work into machine production.' (However, Section IV of Chapter XIV of Volume III does not deal, nor is it intended to deal, with the general effects of relative overpopulation on wages and the degree of exploitation of labour.)[15] This restriction also applies to the subsequent devaluation of constant capital,[16] as well as to all methods of production 'which

But, notably, it is the prolongation of the working day, this invention of modern industry, which increases the mass of appropriated surplus labour without essentially altering the proportion of the employed labour-power to the constant capital set in motion by it, and which rather tends to reduce this capital relatively.' (*ibid*. pp.232-33.)

[15] Sweezy is therefore wrong when he raises the following objection to the argument of this section: 'It would seem, however, that a more important effect of the reserve army . . . is through competition on the labour market with the active labour force, to depress the rate of wages and in this way to elevate the rate of surplus-value.' (*Theory of Capitalist Development*, p.99.) Without a doubt, if Marx had discussed this subject in Chapter XIV this would have led, in fact, to what Bortkiewicz criticised him for.

[16] The fact that Marx included the devaluation of constant capital in the tendencies which counteract the fall in the rate of profit arouses Sweezy's disapproval. He says: 'It might seem that it would be preferable to look first at what might be called the "original" increase of the organic composition, to observe the effects of this on the rate of profit, and only then to take account of the cheapening of the elements of constant capital which is itself due to the rise in productivity associated with the "original" increase. It might be held that if this were done, the rate of increase of the organic composition would appear much larger and that this fact is prevented from showing in the statistics only by one of the "counteracting causes". It is doubtful, however, whether any useful purpose can be served by such an attempt to preserve Marx's implied distinction between the primary rise in the organic composition and the counteracting (but smaller) fall due to the cheapening of the elements of constant capital. All that can be observed is the net change in the organic composition which is the resultant of both forces. It seems better, therefore, to use the expression "change in the organic composition of capital" only in the net sense which takes account of cheapening of the elements of constant capital. If this is done there will perhaps be less temptation to think of the organic composition in physical instead of value terms.' (*ibid*. pp.103-04.)

If we are to believe Sweezy, Marx takes the technical composition of capital as the basis of the law so that he can smuggle in the value composition as a 'counteracting influence'.

raise the rate of profit at a constant rate of surplus-value, or independently of the rate of surplus-value'.[17] And if the critics fail to notice the methodological distinction between Chapters XIII and XIV, this probably has less to do with the rather complex structure of these chapters, than with the preconceived ideas with which they set about their study.

IV. However, if Marx himself did take account of the mutual relation between the organic composition and the rate of surplus-value, i.e. if his law is not based on an arbitrary 'method of isolation', aren't we then compelled to accept an interpretation which denies the tendency of the rate of profit to fall? Wouldn't we then have every reason to agree with Sweezy that : 'If both the organic composition of capital and the rate of surplus-value are assumed variable . . . then the direction in which the rate of profit will change becomes indeterminate. All we can say is that the rate of profit will fall if the percentage increase in the rate of surplus-value is less than the percentage decrease in the proportion of variable to total capital.' However, 'there is no general presumption that changes in the organic composition of capital will be relatively so much greater than changes in the rate of surplus-value that the former will dominate movements in the rate of profit. On the contrary, it would seem that we must regard the two variables as of roughly co-ordinate importance . . . In the general case, therefore, we ought to assume that the increasing organic composition of capital proceeds *pari passu* with a rising rate of surplus-value.'[18] Or, as Natalie Moszkowska says : 'The rate of profit would only fall with technical progress, if it were only the composition of capital which rose and not labour productivity as well. If the rising productivity of labour causes the value of the material and human means of production to fall, then the composition of capital (c/v) will subsequently fall and the rate of surplus-value (s/v) rise. Immediately after the introduction of a technical innovation, equipping the workers with expensive means of production does indeed raise the composition of capital; but after the cheapening of the means of production due to rising productivity of labour, it falls again. And since wages also fall after cheapening of the workers' consumer goods, i.e. the rate of surplus-value rises, then the rate of profit cannot fall.'[19]

[17] *Capital* III, p.236.

[18] Sweezy, *Theory of Capitalist Development*, pp.102, 104.

[19] Moszkowska, *Zur Kritik moderner Krisentheorien*, 1935, p.46. In her previous book (*Das Marxsche System*, 1929, p.118) she wrote : 'The "law of the tendency of the rate of profit to fall" is not a historical, but a dynamic law.

Marx's critics consequently stake their arguments on factors which counteract the tendency of the rate of profit to fall – on the one hand, the subsequent devaluation of the elements of constant capital, and, on the other, the increase in the rate of surplus-value. No one can deny that these factors are at work : the key question is, to what extent do they assert themselves?

As far as the first factor is concerned, it is sufficient in this context to look at one passage in Part III of *Theories*, the chapter on Cherbuliez, which the critics did not notice. We read there : 'There can be no doubt that machinery becomes cheaper, and this for two reasons : (1) the application of machinery to the production of raw materials from which the machinery is made. (2) The application of machinery in the transformation of these materials into machinery. In saying this, we already say two things. Firstly, that in both these branches, compared with the instruments required in the manufacturing industry, the value of the capital laid out in machinery also grows as compared with that laid out in wages. Secondly, what becomes cheaper is the individual machine and its component parts, but a system of machinery develops : the tool is not simply replaced by a single machine, but by a whole system, and the tools which perhaps played the major part previously . . . are now assembled in thousands. Each individual machine confronting the worker is in itself a colossal assembly of instruments which he formerly used singly, e.g. 1,800 spindles instead of one. But, in addition, the machine contains elements which the old instrument did not have. Despite the cheapening of individual elements, the price of the whole aggregate increases enormously and the productivity consists in the continuous expansion of the machinery.' Marx continues : 'It is therefore self-evident or a tautological proposition that the increasing productivity of labour caused by machinery corresponds to increased value of the machinery relative to the amount of labour employed (consequently to the value of labour), the variable capital.'[20]

But what about the raw materials? 'It is obvious that the quan-

It does not confirm a historical fact, namely that the rate of profit falls; it only formulates the mutual dependence of two variables, namely 1. If the rate of surplus-value remains constant, the profit rate falls. 2. If the profit rate remains constant, the rate of surplus-value increases.

That is, the law simply expresses a functional connection. And for this reason one could equally call the law of the "tendency of the rate of profit to fall" the law of the "tendency of the rate of surplus-value to rise".' With the same logic, Moszkowska could call it the 'law of the falling, or not-falling rate of profit'. It is clear that such an interpretation would destroy Marx's law.

[20] *Theories* III, pp.366-67.

tity of raw material must increase proportionally with the productivity of labour; that is, the amount of raw material must be proportionate to that of labour. [But can't this growth in the amount be made up for by a growth in productivity, which reduces the value to exactly the same amount?]²¹ . . . If, for example, productivity in spinning increases tenfold, that is, a single worker spins as much as ten did previously, why should not one Negro produce ten times as much cotton as ten did previously, that is why should the *value* ratio not remain the same? The spinner uses ten times as much cotton in the same time. The ten times larger amount of cotton therefore costs no more than a tenth of this amount cost previously. This means that despite the increase in the amount of the raw material, its value ratio to variable capital remains the same . . . To this it is quite easy to answer that some kinds of raw materials, such as wool, silk, leather, are produced by animal organic processes, while cotton, linen etc. are produced by vegetable organic processes and capitalist production has not yet succeeded, and never will succeed in mastering these processes in the same way as it has mastered purely mechanical or inorganic chemical processes. Raw materials such as skins etc. and other animal products become dearer partly because the . . . law of rent increases the value of these products as civilisation advances. As far as coal and metal . . . are concerned, they become much cheaper with the advance of production; this will however become more difficult as mines etc. are exhausted.' Marx concludes : 'The cheapening of raw materials, and of auxiliary materials, etc. checks but does not cancel the growth in the value of this part of capital. It checks it to the degree that it brings about a fall in profit. This rubbish is herewith disposed of.'²²

V. Finally, how should we judge the main counter-argument offered by Marx's critics – namely, as they put it, the increase in the rate of surplus-value *pari passu* with the rise in organic composition? This argument overlooks several details. First, that 'the value of labour-power does not fall in the same degree as the productivity of labour or of capital increases'. This is because 'this increase in productive power likewise increases the ratio between constant and variable capital in all branches of industry which do not produce necessaries (either directly or indirectly) *without giving rise to any kind of alteration in the value of labour*. The development of productive power is not uniform. It is in the nature of capitalist production that it devel-

²¹ This sentence in brackets was added by Kautsky, the original editor of *Theories*.
²² *Theories* III, pp.367-69.

ops industry more rapidly than agriculture.[23] This is not due to the
nature of the land, but to the fact that in order to be exploited really
in accordance with its nature land requires different social relations
. . . An additional factor is that, as a consequence of landownership,
agricultural products are expensive compared with other commodities, because they are sold at their value and are not forced down to
the price of production. They form, however, the principal constituent of the means of subsistence.[24] Furthermore, if one-tenth of the
land is dearer to exploit than the other nine-tenths, these latter are
likewise hit "artificially" by this relative barrenness, as a result of the
law of competition.'[25] On the other hand, however, 'impeding factors'
enter here, such as, for example, 'that the workers themselves,
although they cannot prevent reductions in (real) wages, will not
permit them to be reduced to the absolute minimum; on the contrary
they achieve a certain quantitative participation in the general
growth of wealth'.[26]

This is not the only reason for the mistakes made by Marx's
critics! Much more important is the fact that they also overlook that
the increase in the rate of profit secured by raising the intensity of
the exploitation of labour is no abstract procedure or arithmetical
operation; rather, it is always related to actual living workers,
and their performance. In other words, the surplus labour which a
worker can perform has definite limits. On the one hand the length
of the working day, and on the other the part of the working day
necessary for the reproduction of the workers themselves. For
example, if the normal working day amounts to 8 hours, no increase
in productive power can squeeze more surplus labour out of the
worker than 8 minus as many hours as correspond to the production
of the wage. If the technique of production succeeded in reducing
the necessary labour-time from e.g. 4 hours to half an hour, then
surplus labour would still not come to more than 15/16 of the work-

[23] The extent to which the differences between industry and agriculture
can be reduced through technical progress is clearly a question which cannot
be discussed here.

[24] Once more, a reservation is necessary here. The North American
worker spends a much smaller portion of his or her wage on food than the
European worker, who in turn spends less than the Asian.

[25] *Theories* III, pp.300-01. As one can see, Marx expressed himself on
this point as clearly and specifically as possible. Despite this, one still reads in
Joan Robinson that 'it might be argued that Marx was unconsciously assuming
that increasing productivity does not affect the wage-good industries, so that
constant real wages are compatible with a constant rate of exploitation.' (*Essay
on Marxian Economics*, p.40.) This passage contains as many mistakes as
words.

[26] *Theories* III, p.312.

ing day (with an 8-hour day); it would increase from 4 to 7½; i.e. not even double. At the same time the productivity of labour would have to grow enormously (as Marx pointed out in the *Rough Draft*)! 'The larger the surplus-value of capital *before the increase of productive force*, the larger the amount of presupposed surplus labour, or surplus-value of capital; or the smaller the fractional part of the working day which forms the equivalent of the worker, which expresses necessary labour, the smaller is the increase in surplus-value which capital obtains from the increase of productive force. Its surplus-value rises, but in an ever smaller relation to the development of the productive force. Thus the more developed capital already is . . . the more terribly must it develop the productive force in order to valorise itself, i.e. to add surplus-value to itself, even to a slight degree – because its barrier always remains the relation between the fractional part of the day which expresses *necessary labour*, and the entire working day. It can move only within these boundaries.'[27] We should not forget that the increase in productivity is not simply expressed in the decline of variable capital (the paid part of the working day) relative to constant, but also in the 'decline of the entire living labour applied to the means of production . . . in relation to the value of these means of production'. I.e. the relation $(v + s) : c$ also falls. This is not, however, essentially a question of a change in the value composition of capital, but is rather related to the change in its technical composition, which characterises technical progress. Thus, if, given a constant capital of definite size, 20 workers were reduced to 10; and if previously the 20 workers worked 80 hours for themselves and 80 for the employer, then it would be impossible for the remaining 10 workers to perform as much surplus labour as the 20 did before, since their entire working time only amounts to 80 hours.[28]

[27] We have already cited these arguments in Chapter 16. See also *Grundrisse*, p.340.

[28] Cf. *Capital* III, p.247: 'Inasmuch as the development of the productive forces reduces the paid portion of employed labour, it raises the surplus-value, because it raises its rate; but inasmuch as it reduces the total mass of labour employed by a given capital, it reduces the factor of the number by which the rate of surplus-value is multiplied to obtain its mass. Two workers, each working 12 hours a day, cannot produce the same mass of surplus-value as 24 who only work two hours, even if they could live on air and hence did not have to work for themselves at all. In this respect, then, the compensation of the reduced number of workers by intensifying the degree of exploitation has certain insurmountable limits. It may, for this reason, well check the fall in the rate of profit, but it cannot prevent it altogether.' Joan Robinson even succeeds in misinterpreting this passage, which she quotes word

In order to maintain the level of surplus labour the working day must either be extended or the intensity of work increased. (In fact, the employer can employ more workers with an enlarged capital, and thus compensate for the fall in the rate of profit by an increase in the mass of profit. But this is quite another question.) All this was fully understood by Ricardo's 'proletarian opponents' (Marx's expression) 120 years ago, when Thomas Hodgskin and the author of the text *The Source and Remedy of the National Difficulties* (1821) derived the fall in the rate of profit from the impossibility of the unlimited extension of surplus labour (or, as they expressed it, the inability of the worker to satisfy the overwhelming demands of 'compound interest'.) Commenting on their views in the *Theories* Marx says that, in the long run, the growth of surplus labour could only compensate for the proportional decline in the labour employed if the working day were to be 'infinitely extended', or necessary labour reduced to zero,[29] both of which appear equally absurd.[30]

We thus come back to the law formulated in Volume I of *Capital*. '*The absolute limit of the average working day* – this being by nature always less than 24 hours – sets an *absolute limit to the compensation for a reduction of variable capital by a higher rate of surplus-value, or for the decrease of the number of workers exploited by a higher degree of exploitation of labour-power*.' 'This self-evident

for word in the following commentary: 'Productivity may rise without limit, and, if real wages are constant, the rate of exploitation rises with it. Marx appears to have been in some confusion upon this point, for when he begins to discuss the effect of a rise in productivity on the rate of exploitation, he switches over in the middle of the argument to discussing the effect of changing the length of the working day.' (*op. cit.* p.39.)

[29] *Theories* III, p.312.

[30] Robinson seems now to interpret the problem in this sense when she writes: 'The trouble probably arose, like most of the obscurities in Marx's argument, from his method of reckoning in terms of value. With given labour-time, of given intensity, the rate of value created is constant. Thus v + c is constant. It might seem, at the first glance, that s/v can rise only if wages fall. But this is an illusion. An increase in productivity reduces the value of commodities, and the value of labour-power, with constant real wages. Thus v falls towards zero, and s/v rises towards infinity, and all the time real wages are constant.' (*op. cit.* pp.39-40.) Admittedly, as the working day consists of two parts, necessary and surplus labour, it follows that if necessary labour constantly falls, surplus labour must rise. (Why the 'method of reckoning in values' has to be given up to understand this simple fact is impossible to fathom.) However, this tautology cannot bring about miracles; it cannot alter the fact that the increase in the extent of exploitation of labour can only make up within definite, narrow limits for the loss of actually performed surplus labour, the loss which arises through the continual reduction in the number of workers employed per unit of capital.

law', Marx adds, 'is of importance for the explanation of many
phenomena, arising from a tendency of capital to reduce as much
as possible the number of workers employed i.e. the amount of its
variable component, the part which is changed into labour-power,
which stands in contradiction with its other tendency to produce
the greatest possible *mass of surplus-value*.'[31] This is clearly an
allusion which prepares us for the solution of the 'riddle of the falling
rate of profit' contained in Volume III of *Capital* – but once again,
this was not noticed by the critics.

Von Bortkiewicz enjoys great popularity in the Anglo-American
school of marxist economics (Sweezy and Meek among others), not
so much because of his objections to Marx's law of the tendency of
the rate of profit to fall, but rather his critique of Marx's 'transformation of values into prices of production' (*Capital* III, Chapter IX).
To us this second aspect of Bortkiewicz's critique seems even less satisfactory than the first, and we regard it as a mere 'academic whim'.
Bortkiewicz's supporters propose the thesis that 'Marx's method of
transformation would lead to a violation of the equilibrium of simple
reproduction', and is therefore 'logically unsatisfactory'.[32] However,
this objection would only be valid if Marx were in fact a 'Harmonist',
i.e. if his schemes of reproduction were to be interpreted in the way
adopted by Tugan-Baranovsky. (It is self-evident that the transition
from commodity-values to 'prices of production' would necessarily be
accompanied by disturbances in the 'equilibrium of simple reproduction'; but since when has it been the task of marxists to prove that
it is theoretically possible for the capitalist economy to proceed without disturbances?) There is also a second point: von Bortkiewicz's
supporters overlook the fact that Marx's 'prices of production' are not
in fact 'prices' at all, but simply values modified by the intervention
of the average rate of profit, and so the 'price calculation' suggested
by Bortkiewicz cannot make the slightest contribution towards solving
the question of the actual 'transformation of values into prices'. Marx
had already set out the way in which the actual transition from values
to prices occur both in the *Grundrisse* and the *Critique*, and it is
superfluous to look for a surrogate solution to this problem.

[31] *Capital* I, pp.419-20 (305-06). Cf. *Capital* III, p.398: 'The identity of
surplus-value and surplus labour imposes a qualitative limit upon the accumulation of capital. This consists of the total working day, and the prevailing
development of the productive forces and of the population, which limits the
number of simultaneously exploitable working days. But if one conceives of
surplus-value in the meaningless form of interest, the limit is merely quantitative and defies all imagination.'
[32] Paul Sweezy, *Theory of Capitalist Development*, pp.88-89.

PART SIX
Conclusion

28.
The Historical Limits of the Law of Value. Marx on the Subject of Socialist Society

According to Marx's original outline the last book of his work was to conclude with an examination of those moments which 'point beyond the presupposition' and 'drive towards the adoption of a new historical form'. That is, it was intended to deal with the 'dissolution of the mode of production based on exchange-value' and its transition to socialism.[1] The central focus of this was naturally the question of the destiny of the law of value, and it is to this subject that we now turn our attention.

1. Marx on the development of human individuality under capitalism

It is well known that the founders of marxism rejected all forms of 'depicting the future', inasmuch as this involved the construction of completed socialist systems, to be derived from the 'eternal principles of justice' and the 'immutable laws of human nature'. Although such systems may have been necessary and justified in those periods when they were first expounded, they developed into a hindrance to the growing workers' movement as soon as it had

[1] *Grundrisse*, pp.228, 264.

acquired a scientific basis in the materialist interpretation of history, established by Marx and Engels. This basis was far superior to the doctrines of the utopian socialists and had to pose the question of a socialist conception of the future in a radically different way. Social- ism no longer appeared as a mere ideal, but as a necessary phase in the development of humanity, the real culmination of all previous history; consequently the socialist society of the future could only be spoken of inasmuch as visible seeds of this new social form could be discovered in history and its developmental tendencies. This does not of course mean that Marx and Engels had no conception of the socialist economic and social order (a view often attributed to them by opportunists), or that they simply left the entire matter to our grandchildren, as if this constituted the scientific character of their theories. On the contrary, such conceptions played a prominent role in Marx's theoretical system. This can be convincingly shown if we examine the main works of the founders of marxism. Take, for example, Marx's *Capital*, which sought both to investigate the inner structure and laws of motion of the capitalist mode of production *and* to produce the proof and necessity of the 'great change' which was to bring about the abolition of human 'self-alienation', through which humanity would become the 'real conscious rulers of nature and their own association' (Engels). We therefore constantly encounter discussions and remarks in *Capital*, and the works preparatory to it, which are concerned with the problems of a socialist society, and which make especially clear both what Marx had in common with the theories of the utopian socialists, and where he differed from them.

These discussions were a necessary product of Marx's dialectical and material method, which sought to grasp every social phenomenon in the flow of its becoming, existence and passing away. This method therefore directs our attention both to 'earlier historical modes of production'[2] and looks forward to those points 'at which the sus- pension of the present form of production relations gives signs of its becoming – foreshadowing the future. Just as, on the one side, the pre-bourgeois phases appear as *merely historical*, i.e. suspended pre- suppositions, so do the contemporary conditions of production like- wise appear as engaged in *suspending themselves* and hence in posit- ing the *historic presuppositions* for a new state of society.'[3]

Thus the dialectical-materialist study of the capitalist mode of production leads directly to counterposing this mode of production,

[2] See Note 3 on p.268 above.
[3] *Grundrisse*, p.461.

on the one hand, to pre-capitalist social formations, and on the other, to the socialist social order which replaces it. 'The *private exchange* of all products of labour, all activities and all wealth stands in antithesis not only to a distribution based on a natural or political super- and subordination of individuals to one another . . . (regardless of the character of this super- and subordination; patriarchal, ancient or feudal) but also to free exchange among individuals who are associated on the basis of common appropriation and control of the means of production.'[4] This yields a division of the whole of human history into three stages, in the form of a dialectical triad : 'Relations of personal dependence (entirely spontaneous at the outset) are the first social forms, in which human productive capacity develops only to a slight extent and at isolated points. Personal independence founded on *objective* dependencies is the second great form, in which a system of general social metabolism, of universal relations, of all-round needs and universal capacities is formed for the first time. Free individuality, based on the universal development of individuals and on their subordination of their communal, social productivity as their social wealth, is the third stage. The second stage creates the conditions for the third.'[5]

Human history is therefore seen in terms of its most basic final outcome; as a necessary process of the elaboration and development of the human personality and its freedom. However, from Marx's point of view the issue was not so much to demonstrate the necessity of this process (this was already recognised by classical German philosophy), but rather to liberate this discovery from ideological illusions and place it on the firm foundation of real history i.e. the development of the social relations of production. This task could only be accomplished with the assistance of the materialist method.

The *Rough Draft* states : 'When we look at social relations which create an undeveloped system of exchange, of exchange-values and of money [i.e. pre-capitalist relations] . . . then it is clear from the outset that the individuals in such a society, although their relations appear to be more personal, enter into connection with one another only as individuals imprisoned within a certain definition, as feudal lord and vassal, landlord and serf etc. or as members of a caste etc. or as members of an estate etc. In the money relation, in the developed system of exchange (and this semblance seduces the democrats), the ties of personal dependence, of distinctions of blood, education, etc. are in fact exploded, ripped up . . . and individuals

[4] *ibid.* p.159.
[5] *ibid.* p.158.

seem independent[6] . . . free to collide with one another and to engage in exchange within this freedom; but they appear thus only for someone who abstracts from the *conditions*, the conditions of existence within which these individuals enter into contact . . . The definedness of individuals, which in the former case appears as a personal restriction of the individual by another, appears in the latter case as developed into an objective restriction of the individual by relations independent of him and sufficient unto themselves. Since the single individual cannot strip away his personal definition, but may well overcome and master external relations, his freedom *seems* to be greater in case 2. A closer examination of these external relations, these conditions, shows, however, that it is impossible for the individuals of a class etc. to overcome them *en masse* without destroying them. A particular individual may by chance get on top of these relations, but the mass of those under their rule cannot, since their mere existence expresses subordination, the necessary subordination of the mass of individuals. These external relations are very far from being an abolition of "relations of dependence" : they are rather the dissolution of these relations into a general form; they are merely the elaboration and emergence of the general *foundation* of the relations of personal dependence.'[7]

We read in a marginal note to the examination of the 'objective power of money' in the *Rough Draft* : 'It has been said and may be

---

[6] Marx adds: 'This is an independence which is at bottom merely an illusion, and it is more correctly called indifference.'

[7] *Grundrisse*, pp.163-64. We read further on in the text: 'These objective dependency relations also appear, in antithesis to those of personal dependence (the objective dependency relation is nothing more than social relations which have become independent and now enter into opposition to the seemingly independent individuals; i.e. the reciprocal relations of production separated from and autonomous of individuals) in such a way that individuals are now ruled by abstractions, whereas earlier they depended on one another. The abstraction, or idea, however, is nothing more than the theoretical expression of those material relations which are their lord and master. Relations can be expressed, of course, only in ideas, and thus philosophers have determined the reign of ideas to be the peculiarity of the new age, and have identified the creation of free individuality with the overthrow of this reign. This error was all the more easily committed from the ideological standpoint, as this reign exercised by the relations (this objective dependency, which incidentally, turns into definite relations of personal dependency, but stripped of all illusions) appears within the consciousness of individuals as the reign of ideas, and because the belief in the permanence of these ideas, i.e. of these objective relations of dependency, is of course consolidated, nourished and inculcated by the ruling classes by all means available.' (*ibid.* pp.164-65.) Cf. *German Ideology*, pp.49ff.

said that this is precisely the beauty and greatness of it : this spontaneous interconnection, this material and mental metabolism which is independent of the knowing and willing of individuals, and which presupposes their reciprocal independence and indifference. And certainly this objective connection is preferable to the lack of any connection, or to a merely local connection resting on blood ties, or on primeval, natural or master-servant relations. Equally certain is it that individuals cannot gain mastery over their own social interconnections before they have created them.[8] But it is an insipid notion to conceive of this merely *objective bond* as a spontaneous, natural attribute inherent in individuals and inseparable from their nature . . . This bond is their product. It is a historic product. It belongs to a specific phase of their development. The alien and independent character in which it presently exists *vis-à-vis* individuals proves only that the latter are still engaged in the creation of the conditions of their social life, and that they have not yet begun, on the basis of these conditions, to live it. It is the bond of individuals within specific and limited relations of production.' However : 'In earlier stages of development the single individual seems to be developed more fully, because he has not yet worked out his relationships in their fullness, or erected them as independent social powers and relations opposite himself. It is as ridiculous to yearn for a return to that original fullness as it is to believe that with this complete emptiness', which characterises the 'new age',[9] history has come to a standstill.[10] The bourgeois viewpoint has never advanced beyond this anti-

[8] i.e. cannot make the transition to a socialist society.

[9] Marx writes in similar vein in *Capital* on the subject of the modern machine operator : 'Even the lightening of the labour becomes an instrument of torture, since the machine does not free the worker from the work, but rather deprives the work itself of all content . . . The special skill of each individual machine operator who has now been deprived of all significance vanishes as an infinitesimal quantity in the face of the science, the gigantic natural forces, and the mass of social labour embodied in the system of machinery, which, together with those three forces, constitutes the power of the "master".' *Capital* I, pp.548-49 (423).

[10] Cf. *Contribution*, p.95 : 'So little does the relation of buyer and seller represent a purely individual relationship that they enter into it only insofar as their individual labour is negated, that is to say, turned into money as non-individual labour. It is therefore as absurd to regard buyer and seller, these bourgeois economic types, as eternal social forms of human individuality, as it is preposterous to weep over them as signifying the abolition of individuality.'

It is interesting to note that a similar passage can be found in the works of the young Hegel. In his text *The German Constitution* (1798-99), which survives only as fragments, he writes on the subject of early pre-constitutional

thesis between itself and this romantic viewpoint, and therefore the latter will accompany it as legitimate antithesis up to its blessed end.'[11]

The main defect with the bourgeois concept of freedom is now clear; it is the unhistorical approach of its advocates, who regard the development of individuality, which is specific to one particular epoch and one particular mode of production, as being absolute and confuse this with the realisation of 'freedom *tout court*'. ('Just like a man who believes in a particular religion and sees it as *the* religion, and everything outside it as *false* religions.'[12]) They simply fail to grasp that bourgeois freedom – far from representing the embodiment of 'freedom in general', is rather a highly specific product of the capitalist mode of production, and consequently shares all its limitations. For human beings, having been liberated from earlier limitations, are subject to a new fetter under capitalism, namely the reified rule of their relations of production, which have grown up over them, the blind power of competition and chance.[13] In one respect they have become freer, but in another they have become less free.

This unhistorical mode of thinking is shown most clearly in the attitude of bourgeois economists (and bourgeois ideologists in general) towards capitalist competition. Marx says that although competition 'appears historically as the dissolution of compulsory guild membership, government regulation, internal tariffs and the like within a country, as the lifting of blockades, prohibitions, and protection on the world market', it has never 'been examined for this merely negative side . . . its merely historical side; . . . and this has led at the same time to the even greater absurdity of regarding it as the collision of unfettered individuals who are determined only by their own interests – as the mutual repulsion and attraction of free individuals, and hence as the absolute mode of existence of free individuality in the sphere of consumption and of exchange.'

'German freedom' : 'Just as it is cowardly and effete to describe the men of that society as loathsome, wretched and stupid, and to imagine ourselves to be infinitely more human, happy and clever, so too it is childish and silly to yearn for such a society – as if it alone were natural – or to fail to recognise that a society ruled by law is necessary – and that it alone is free.' (Quoted in Lukacs, *The Young Hegel*, p.141.)

[11] *Grundrisse*, pp.161-62.

[12] *Theories* II, p.529.

[13] Cf. Marx's 'Saint Max' (Stirner): 'It has already been pointed out to him that in competition personality itself is a matter of chance, while chance is personality.' (*German Ideology*, p.421.)

'Nothing can be more mistaken,' Marx adds. For, in the first place, 'while free competition has dissolved the barriers of earlier relations and modes of production, it is necessary to observe first of all that the things which were a barrier to it were the inherent limits of the earlier mode of production, within which they spontaneously developed and moved. The limits became barriers only after the forces of production and the relations of intercourse had developed sufficiently to enable capital as such to emerge as the dominant principle of production. The limits which it tore down were barriers to its motion, its development and realisation. It is by no means the case that it thereby suspended all limits, nor all barriers, but rather only the limits not corresponding to it, which were barriers to it.[14] Within its own limits – however much they may appear as barriers from a higher standpoint . . . it feels free, and free of barriers, i.e. as limited only by itself, only by its own conditions of life. Exactly as guild industry, in its heyday, found in the guild organisation all the fullness of freedom it required i.e. the relations of production corresponding to it. After all, it posited these out of itself, and developed them as its inherent conditions, and hence in no way as external and constricting barriers. The historical side of the negation of the guild system etc. by capital through free competition signifies nothing more than that capital, having become sufficiently strong, by means of the mode of intercourse adequate to itself, tore down the historic barriers which hindered and blocked the movement adequate to it.'

Nevertheless, competition has by no means only this negative, merely historical significance; it is, in essence, the realisation of the capitalist mode of production![15] Thus, the statement that 'within free competition the individuals, in following purely their private interest, realise the communal or rather the *general*[16] interest' simply expresses an illusion. For 'it is not individuals who are set free by free competition; it is rather capital which is set free. As long as production resting on capital is the necessary, hence the fittest form for the develop-

[14] This instance (the mutual relation between 'limit' and 'barrier') also illustrates the use of Hegelian concepts.

[15] Cf. p.44 above.

[16] In Marx's terminology (this applies especially to the young Marx) the 'general' is no way identical to the 'communal', but rather refers to what arises from the collision of 'communal' and 'particular interests', in a society of atomised, private owners. (Cf. *German Ideology*, p.46: 'Just because individuals seek only their particular interest, which for them does not coincide with their communal interest [in fact the general is the illusory form of communal life], the latter will be imposed on them as an interest "alien" to them, and independent of them, as in its turn a particular peculiar "general interest".')

ment of the force of social production, the movement of individuals within the pure conditions of capital appears as their freedom; which is then also dogmatically propounded as such through constant reflection back on the barriers torn down by free competition.'[17] Hence 'the insipidity of the view that free competition is the ultimate development of human freedom : and that the negation of free competition = negation of individual freedom and of social production based on individual freedom. It is nothing more than free development on a limited basis – the basis of the rule of capital. This kind of individual freedom is therefore at the same time the most complete abolition of individual freedom, and the most complete subjugation of individuality under social conditions which assume the form of objective powers, even of overpowering objects. The analysis of what free competition really is, is the only rational reply to the middle-class prophets who laud it to the skies or to the socialists who damn it to hell.'[18] In fact 'the assertion that free competition = the ultimate form of the development of the forces of production and hence of human freedom means nothing other than that middle-class rule is the culmination of world history – certainly an agreeable thought for the parvenus of the day before yesterday.'[19]

It is clear that what we are reading here is simply a continuation of ideas previously encountered in the *German Ideology*; namely, that in the course of human history the development of the productive forces has created a situation in which the original, personal relations of dependence have been replaced by simple objective ones, and where the local and national social connection between people has been replaced by a universal connection. In the *German Ideology* Marx and Engels already pointed to the contradictory and two-sided character of previous social progress, which on the one hand gave rise to the creation of a social individual, more capable of development, and with abundant needs, but on the other hand produced the most extensive 'alienation' and 'emptying' of this individual. And finally we also find the argument according to which the liberation of mankind from feudal and other constraints, by capitalism, produces an apparent freedom only, and complete freedom, the 'original and free development of individuals' will only become a

---

[17] Marx adds: 'By the way, when the illusion about competition as the so-called absolute form of free individuality vanishes, this is evidence that the conditions of competition, i.e. of production founded on capital, are already felt and thought of as barriers, and hence already are such, and more and more become such.'

[18] That is, the Proudhonists.

[19] *Grundrisse*, pp.649-52.

reality under communism. We read in the *German Ideology* : 'In imagination individuals seem freer under the dominance of the bourgeoisie than before, because their conditions of life seem accidental; in reality of course, they are less free, because they are more subjected to the violence of things.' And it was just 'this right to the undisturbed enjoyment, within certain conditions, of fortuity and chance [which] has up to now been called personal freedom'.[20] This conception is developed further in the *Rough Draft*; except that there the alternative, positive aspect of the contradiction – the actual progress which 'apparent bourgeois freedom' brings about – is expressed much more clearly and emphatically.

This is best seen in the remarkable passage dealing with the 'childish world of antiquity' in antithesis to the modern world of capitalism. Marx says there : 'Do we never find in antiquity an inquiry into which form of landed property etc. is the most productive, creates the greatest wealth? Wealth does not appear as the aim of production, although Cato may well investigate which manner of cultivating a field brings the greatest rewards, and Brutus may even lend out his money at the best rates of interest. The question is always which mode of property creates the *best citizens*.' The situation is completely different in the modern world. Here wealth 'appears in all forms in the shape of a thing, be it an object or be it a relation mediated through the object, which is external and accidental to the individual. Thus the old view, in which the human being appears as the aim of production, regardless of his limited national, religious, political character, seems to be very lofty when contrasted to the modern world, where production appears as the aim of mankind and wealth as the aim of production. In fact, however, when the limited bourgeois form is stripped away, what is wealth other than the universality of individual needs, capacities, pleasures, productive forces etc., created through universal exchange? The full development of human mastery over the forces of nature, those of so-called nature as well as of humanity's own nature? The absolute working-out of his creative potentialities, with no presupposition other than previous historic development, which makes this totality of development, i.e. the development of all human powers as such the end in itself, not as measured on a *predetermined* yardstick? Where he does not reproduce himself in one specificity, but produces his totality? Strives not to remain something he has become, but is in the absolute movement of becoming? In bourgeois economics – and in the epoch of production to which it

[20] *German Ideology*, pp.95, 94.

corresponds – this complete working-out of the human content appears as a complete emptying out, this universal objectification as total alienation,[21] and the tearing down of all limited, one-sided aims as sacrifice of the human end-in-itself to an entirely external end. This is why the childish world of antiquity appears on the one side as loftier. On the other side, it really is loftier in all matters where closed shapes, forms and given limits are sought for. It is satisfaction from a limited standpoint; while the modern age gives no satisfaction; or, where it appears satisfied with itself, it is *vulgar.*'[22]

The antithesis between the marxist and the Romantic critique of capitalism is expressed here with particular clarity – Marx did not just attack the Romantics because of their 'sentimental tears',[23] or because they 'waved the alms-bag of the proletariat in front for a banner', for demagogic reasons, while at the same time hiding the 'old feudal coat-of-arms' behind their backs.[24] He reproached them much more with being totally incapable of grasping the 'course of modern history', i.e. the necessity and the historical progressiveness of the bourgeois social order which they criticised, and for confining themselves instead to moralistic rejection of it.

No one would deny that the rule of capital is based on the most callous extraction of surplus labour, on the exploitation and oppression of the mass of the people. In this respect, it certainly 'surpasses all previous systems of production, which were based on directly compulsory labour, in its energy and its quality of unbounded and ruthless activity'.[25] But only capital has 'subjugated historical progress to the service of wealth';[26] it is only the capitalist form of production which 'becomes an epoch-making mode of exploitation, which, in

---

[21] 'What, then, constitutes the alienation of labour? First, the fact that labour is external to the worker, i.e. it does not belong to his essential being; that in his work, therefore, he does not affirm himself but denies himself, does not feel content, but unhappy, does not develop freely his physical and mental energy but mortifies his body and ruins his mind. The worker therefore only feels himself outside his work, and in his work feels outside himself. He is at home when he is not working, and when he is working he is not at home.' This state of affairs produces an inversion of all human values in capitalism. 'What is animal becomes human and what is human becomes animal. Certainly eating, drinking, procreating etc. are also genuinely human functions. But abstractly taken, separated from the sphere of other human activity and turned into sole ultimate ends, they are animal functions.' (*Economic and Philosophical Manuscripts*, pp.110-11.)

[22] *Grundrisse*, pp.487-488.

[23] *Economic and Philosophical Manuscripts*, p.100.

[24] *Communist Manifesto. Selected Works*, p.54.

[25] *Capital* I, p.425 (310).

[26] *Grundrisse*, p.590.

the course of its historical development, revolutionises, through the organisation of the labour process and the enormous improvement of techniques, the entire economic structure of society in a manner eclipsing all former epochs'.[27]

Capitalist production is therefore radically different from all previous modes of production by virtue of its universal character, and its drive to continually revolutionise the material forces of production. If pre-capitalist stages of production were never able to increase labour beyond that required for immediate subsistence, due to their primitive, undeveloped techniques, then the 'great historic aspect of *capital*' consists in the fact that it 'produces surplus labour, surplus from the standpoint of simple use-value, of mere subsistence'; and it carries out this task by developing, on the one hand, the social forces of production, and on the other, human needs and capacities for labour, to an extent that has never existed before.

A particularly striking passage from the *Rough Draft* reads as follows : 'The historic destiny of capital is fulfilled as soon as, on one side, there has been such a development of needs that surplus labour above and beyond necessity has itself become a general need arising out of individual needs themselves – and, on the other, when the severe discipline of capital acting on succeeding generations has developed general industriousness as the general property of the new species';[28] and finally, 'when the development of the productive powers of labour, which capital incessantly whips onwards with its unlimited mania for wealth, and of the sole conditions in which this mania can be realised, have flourished to the stage where the possession and preservation of general wealth require a lesser labour-time of society as a whole, and where the labouring society relates scientifically to the process of its progressive reproduction in a constantly greater abundance; hence where labour in which a human being does what a thing could do for him instead has ceased . . . Capital's ceaseless striving towards the general form of wealth drives labour beyond the limits of its natural paltriness, and thus creates the material elements for the development of the rich individuality which is as all-sided in its production as in its consumption, and whose labour also therefore appears no longer as labour, but as the full

[27] *Capital* II, p.37.
[28] We read in another passage in the *Rough Draft* that capital, 'correctly understood, appears as the condition of the development of the forces of production as long as they require an external spur, which appears at the same time as their bridle. It is a discipline over them, which becomes superfluous and burdensome at a certain level of their development, just like the guild etc.' (*Grundrisse*, p.415.)

development of activity itself,[29] in which natural necessity in its direct form has disappeared; because a historically created need has taken the place of a natural one. This is why *capital is productive*; an *essential relation for the development of the social productive forces*. It ceases to exist as such only where the development of these productive forces themselves encounters its barriers in capital itself.'[30]

In other words, whereas all previous modes of production had to be content with a situation in which the forces of production progressed only very slowly,[31] or even remained stationary over long periods, capital proceeds directly 'from the constant overthrow of its prevailing presuppositions, as the presupposition of its reproduction . . . Although limited by its very nature, it strives towards the universal development of the forces of production, and thus becomes the presupposition of a new mode of production, which is founded not on the development of the forces of production for the purpose of reproducing or at most expanding a given condition, but where the free, unobstructed, progressive and universal development of the forces of production is itself the presupposition of society and hence of its reproduction; where advance beyond the point of departure is the only presupposition.'[32] Only on this new foundation does the 'universality of the individual' become possible, 'not an ideal or imagined universality, but the universality of his real and ideal relations; hence also the grasping of his own history as a *process* and the recognition of nature (equally present as practical power over nature) as his real body.'[33]

Thus, it is the development of capitalism itself which lays the basis for the resolution of the problem of the human personality and human freedom which is posed by history. And seen from this pespective, the historic achievement of capital – which Marx himself pointed to so often and so forcefully – cannot be emphasised enough.

---

[29] 'Labour', says Marx, 'is free in all civilised countries; the point is not to free it, but to abolish it.' (*German Ideology*, p.224.) Cf. Marcuse, *Reason and Revolution* p.293 : 'Marx envisioned the future mode of labour to be so different from the prevailing one that he hesitated to use the same term "labour" to designate alike the material process of capitalist and communist society.'

[30] *Grundrisse*, p.325.

[31] 'All previous forms of society – or, what is the same, of the forces of social production – foundered on the development of wealth. . . . The development of science alone – i.e. the most solid form of wealth both in its product and in its producer – was sufficient to dissolve these communities.' (*ibid.* pp.540-41.)

[32] *ibid.* p.540.

[33] *ibid.* p.542.

2. *The role of machinery as the material precondition of socialist society*

Marx states in the *Rough Draft* : 'If we did not find, concealed in society as it is, the material conditions of production and the corresponding relations of exchange prerequisite for a classless society, then all attempts to explode' existing society, 'would be quixotic'.[34] What are, then, the material conditions of production which render the transition to a classless society both possible and necessary?

The answer is to be found primarily in Marx's analysis of the role of machinery, which shows us, on the one hand, how the development of automatic systems of machinery debases the individual worker to the level of a partial tool, a mere moment in the labour process; but also, on the other hand, how this very same development simultaneously creates the preconditions for the reduction of the expenditure of human energy in the production process to a minimum. And further, that this process permits the replacement of the one-sided worker of today by all-round developed individuals, for whom 'the different social functions are interchangeable modes of activity'. All this can be found in both *Capital* Volume I, and the *Rough Draft*. However, the *Rough Draft* also contains discussions on machinery which are not in *Capital*, and which, despite the fact that they were written more than a hundred years ago, still generate a feeling of awe and excitement, containing as they do some of the boldest visions attained by the human imagination.

Marx writes : 'The exchange of living labour for objectified labour – i.e. the positing of social labour in the form of the contradiction of capital and wage-labour – is the ultimate development of the *value relation* and of production resting on value. Its presupposition is – and remains – the mass of direct labour-time, the quantity of labour employed, as the determinant factor in the production of wealth. But to the degree that large industry develops, the creation of real wealth comes to depend less on labour-time and on the amount of labour employed than on the power of the agencies set in motion during labour-time, whose powerful effectiveness is . . . out of all proportion to the direct labour-time spent on their production, but depends rather on the general state of science and on the progress of technology, or the application of this science to production . . . Real wealth manifests itself rather – and large industry reveals this – in the monstrous disproportion between the labour-time applied, and its product, as well as in the qualitative imbalance between labour,

[34] *ibid.* p.159.

reduced to a pure abstraction, and the power of the production process it superintends. Labour no longer appears so much to be included within the production process; rather, the human being comes more to relate as watchman and regulator to the production process itself . . . No longer does the worker insert a modified natural thing as middle link between the object and himself; rather he inserts the process of nature, transformed into an industrial process, as a means between himself and inorganic nature, mastering it. He steps to the side of the production process instead of being its chief actor. In this transformation, it is neither the direct human labour he himself performs, nor the time during which he works, but rather the appropriation of his own general productive power, his understanding of nature and his mastery over it by virtue of his presence as a social body – it is, in a word, the development of the social individual which appears as the great foundation-stone of production and of wealth. The *theft of alien labour-time, on which the present wealth is based*, appears a miserable foundation in face of this new one, created by large-scale industry itself. As soon as labour in the direct form has ceased to be the great well-spring of wealth, labour-time ceases to be its measure, and hence exchange-value [must cease to be the measure] of use-value. The *surplus labour of the mass* has ceased to be the condition for the development of general wealth, just as the *non-labour of the few*, for the development of the general powers of the human head.[35] With that, production based on exchange-value breaks down, and the direct, material production process is stripped of the form of penury and antithesis. The free development of individualities, and hence not the reduction of necessary labour-time so as to posit surplus labour, but rather the general reduction of the necessary labour of society to a minimum, which then corresponds to the artistic, scientific etc. development of the individuals in the time set free, and with the means created, for all of them.'[36]

Another section from the *Rough Draft* reads : 'The creation of *a large quantity of disposable time* apart from necessary labour-time for society generally and each of its members (i.e. room for the development of the individuals' full productive forces, hence those of society also), this creation of not-labour-time appears in the stage of capital, as of all earlier ones, as not-labour-time, free time, for a few. What capital adds is that it increases the surplus labour-time of the mass by all the means of art and science, because its wealth consists directly in the appropriation of surplus labour-time; since *value*

35 Cf. Chapter 17 above.
36 *Grundrisse*, pp.705-06.

directly its purpose not use-value. It is thus, despite itself, instrumental in creating the means of social disposable time, in order to reduce labour-time for the whole society to a diminishing minimum, and thus to free everyone's time for their own development. But its tendency, always, on the one side, *to create disposable time, on the other to convert it into surplus labour.* If it succeeds too well at the first, then it suffers from surplus production, and then necessary labour is interrupted, because *no surplus labour can be realised by capital.*[37] The more this contradiction develops, the more does it become evident that the growth of the forces of production can no longer be bound up with the appropriation of alien labour, but that the mass of workers must themselves appropriate their own surplus labour. Once they have done so – and *disposable time* thereby ceases to have an *antithetical* existence – then on one side, necessary labour-time will be measured by the needs of the social individual, and, on the other, the development of the power of social production will grow so rapidly that, even though production is now calculated for the wealth of all, *disposable time* will grow for all. For real wealth is the developed productive power of all individuals. The measure of wealth is then not any longer, in any way, labour-time, but rather disposable time. Labour-time *as the measure of value* posits wealth itself as founded on poverty, and disposable time as existing *in and because of the antithesis to surplus labour-time*; or, the positing of an individual's entire time as labour-time, and his degradation therefore to mere worker, subsumption under labour.'[38]

This, then, is Marx's analysis of the historical changes which are brought about by machinery in the capitalist process of production. It is hardly necessary today – in the course of a new industrial revolution – to emphasise the prophetic significance of this enormously dynamic and essentially optimistic conception. For the dreams of the isolated German revolutionary in his exile in London in 1858 have now, for the first time, entered the realm of what is immediately possible. Today, for the first time in history, thanks to the developments of modern technology, the preconditions for a final and complete abolition of the 'theft of alien labour-time' actually exist; furthermore, the present period is the first in which the development

[37] Cf. *Capital* III, pp.255-56: 'Overproduction of capital is never anything more than overproduction of means of production – of means of labour and necessities of life – which may serve as capital, i.e. may serve to exploit labour at a given degree of exploitation; a fall in the intensity of exploitation below a certain point, however, calls forth disturbances, and stoppages in the capitalist production process, crises and destruction of capital.'

[38] *Grundrisse*, p.708.

of the productive forces can be carried so far forward that, in fact, in the not too distant future it will be not labour-time, but rather disposable time, by which social wealth is measured. Whereas previously all methods for increasing the productivity of human labour proved themselves in capitalist practice to be at the same time methods involving increasing degradation, subordination and de-personalisation, today technological development has reached a point where the workers can be finally liberated from the 'serpent of their agonies', from the torture of the assembly-line, from fast and monotonous, subdivided work, and instead of being the appendages of the production process become its real superintendents. There has never been a time when the conditions for a transformation to a socialist society have been so ripe; socialism has never been as indispensable or economically feasible as it is today! One is reminded of the trite bourgeois objection according to which socialism must collapse from the necessity of carrying out hard and unpleasant tasks, which everyone tries to unload onto someone else.[39] Such an objection, based on the nature of the normal bourgeois person, must appear laughable in the face of today's unprecedented development of the productive forces! Naturally, as long as water had to be carried into the house in buckets, there could have been few people who did not try to impose this drudgery on someone else; however, the construction of water-pipes made the profession of water-carrier superfluous. It is now clear that the development of technology works towards a situation where the previous crippling division of labour and its accompanying drudgery disappears, and where, in its place, labour can be posed as the free activity of mental and physical powers. To go back to a witty comparison made by Trotsky:[40] just as it would be stupid for boarders at a good hotel with a plentiful lunch to squabble over butter, bread and sugar, so, in the new society, the exploitation of person by person, the 'theft of alien labour-time' would appear senseless and economically pointless. Not until then will the construction of a really classless, really socialist society, be finally secured.

3. The withering away of the law of value under socialism

And then it is not work as such which will disappear, but the surplus labour of the masses in the interests of, and under the direc-

---

[39] Blanqui had already maliciously remarked that the objection of the bourgeois critics which runs 'Who empties the chamber pots under socialism?' is in fact reducible to the simple question 'Who will empty *my* chamber pot?'
[40] See *The Revolution Betrayed*, p.46.

tives of the few. This is because, stresses Marx, 'labour is the everlasting nature-imposed condition of human existence, and it is therefore independent of every form of that existence, or rather it is common to all forms of society in which human beings live.'[41]

'In the sweat of thy brow shalt thou labour! was Jehovah's curse on Adam. And this is labour for Adam Smith, a curse. "Tranquility" appears as the adequate state, as identical with "freedom" and "happiness". It seems quite far from Smith's mind that the individual, "in his normal state of health, strength, activity, skill, facility",[42] also needs a normal portion of work, and of the suspension of tranquillity. Certainly, labour obtains its measure from the outside, through the aim to be attained and the obstacles to be overcome in attaining it. But Smith has no inkling whatever that this overcoming of obstacles is in itself a liberating activity – and that, further, the external aims become stripped of the semblance of merely external natural urgencies, and become posited as aims which the individual himself posits – hence as self-realisation, objectification of the subject, hence real freedom, his action is, precisely, labour.[43] He is right, of course, that, in its historic forms as slave labour, serf labour, and wage-labour, labour always appears as repulsive, always as *external forced labour*; and not-labour, by contrast, as "freedom, and happiness".'

Marx continues: 'This holds doubly: for this contradictory labour;[44] and, relatedly, for labour which has yet created the subjective and objective conditions . . . in which labour becomes attractive work, the individual's self-realisation, which in no way means that it becomes mere fun, mere amusement, as Fourier, with *grisette*-like *naïveté*, conceives it. Really free working, e.g. composing is at the same time precisely the most damned seriousness, the most intense exertion.'[45] And Marx comes later to speak of Fourier's conception: 'Labour cannot become play as Fourier would like . . . Free

[41] *Capital* I, p.290 (184).

[42] Marx refers here to the following passage from Adam Smith: 'Equal quantities of labour, at all times and places, may be said to be of equal value to the labourer. In his ordinary state of health, strength and spirits; in the ordinary degree of his skill and dexterity, he must always lay down the same portion of his ease, his liberty and his happiness.' (Adam Smith, *Wealth of Nations*, New York, 1937, p.33.)

[43] Cf. *Theories* III, p.257: 'But free time, disposable time, is wealth itself, partly for the enjoyment of the product, partly for free activity which – unlike labour – is not dominated by the pressure of an extraneous purpose which must be fulfilled, and the fulfilment of which is regarded as a natural necessity or a social duty, according to one's inclination.'

[44] i.e. conditioned by a class antagonism.

[45] *Grundrisse*, p.611.

time – which is both idle time and time for higher activity – has naturally transformed its possessor into a different subject, and he then enters into the direct production process as this different subject. This process is then both discipline, as regards the human being in the process of becoming; and, at the same time, practice, experimental science, materially creative and objectifying science, as regards the human being who has become, in whose head exists the accumulated knowledge of society.'[46]

Thus productive human activity, work, will also be of decisive importance in socialist society. Of course, it will undergo dramatic qualitative and quantitative changes! It will be distinguished in qualitative terms from the capitalist form of work, the form which Smith so accurately described as a 'sacrifice of freedom and happiness' by the fact that firstly, the worker will be turned into a conscious director of the production process, whose task will be more and more confined to the simple supervision of the enormous machines and natural forces which operate in production; and secondly, through its character as directly socialised communal work, the product of which will no longer confront the worker in the form of an alien and dominating thing.[47] In this way labour, under socialism, freed from the dross of the past, will lose the repulsive character of forced labour and become 'attractive work' in the sense adopted by Fourier and Owen.[48] This total transformation of work will express itself

[46] *ibid.* p.712.

[47] We read in the *Rough Draft*: 'The emphasis comes to be placed not on the state of being objectified, but on the state of being alienated, dispossessed, sold; on the condition that the monstrous objective power which social labour itself erected opposite itself as one of its moments belongs not to the worker, but to the personified conditions of production, i.e. to capital. To the extent that, from the standpoint of capital and wage-labour, the creation of the objective body of activity happens in antithesis to the immediate labour capacity – that this process of objectification in fact appears as a process of dispossession from the standpoint of capital – to that extent, this twisting and inversion is a real, not merely supposed one existing merely in the imagination of the workers and the capitalists.' But 'the bourgeois economists are so much cooped up within the notions belonging to a specific historic stage of social development that the necessity of the objectification of the powers of social labour appears to them as inseparable from the necessity of their alienation *vis-à-vis* living labour.' (*Grundrisse*, pp.831-32.)

[48] 'It is self-evident', says Marx in the *Theories*, 'that if labour-time is reduced to a normal length and, furthermore, labour is no longer performed for someone else, but for myself, and, at the same time, the social contradictions between master and men etc. are abolished, it acquires a different, free character, it becomes real social labour, and finally the basis of disposable time – the labour of a man who has also disposable time must be of a much higher quality than that of the beast of burden.' (*Theories* III, p.257.)

quantitatively in a fundamental reduction of working-time, and a consequent creation and extension of disposable, free, time. For although even socialist society will not be able to dispense with 'surplus labour' altogether,[49] it will be in a position, thanks to the full unfolding of its productive forces, to reduce the amount of work for individual members of society to a minimum. When this has been accomplished not only will the traditional division of labour, with its separation of people into 'manual' and 'mental' workers, fall to one side, but also the difference between work time and free time will lose its present-day antithetical character, since work time and free time will begin to resemble one another, and complement each other.[50]

Of course, even though labour has been transformed and reduced to a minimum, it will have to be distributed among the various branches of production and individuals, and compared with the desired levels of production; this will require measurement by means of a unitary measure. 'On the basis of communal production, the determination of time remains, of course essential. The less time the society requires to produce wheat, cattle etc. the more time it wins for other production, material or mental. Just as in the case of an individual, the multiplicity of its development, its enjoyment and its activity depends on the economisation of time. Economy of time, to this all economy ultimately reduces itself. Society likewise has to distribute its time correctly in a purposeful way, in order to achieve a production adequate to its overall needs;[51] just as the individual has to distribute his time correctly in order to achieve knowledge in proper proportions or in order to satisfy the various demands on his

[49] It states in *Capital* I, p.667 (530): 'The abolition of the capitalist form of production would permit the reduction of the working day to the necessary labour-time. But even in that case the latter would expand to take up more of the day, and for two reasons: first, because the worker's conditions of life would improve, and his aspirations become greater, and second because a part of what is now surplus labour would then count as necessary labour, namely the labour which is necessary for the formation of a social fund for reserve and accumulation.'

[50] 'It goes without saying, by the way, that direct labour-time itself cannot remain in the abstract antithesis to free time in which it appears from the perspective of bourgeois economy.' (*Grundrisse*, p.712.)

[51] 'It is only where production is under the actual, predetermining control of society that the latter establishes a relation between the volume of social labour-time applied in producing definite articles, and the volume of the social want to be satisfied by these articles.' (*Capital* III, p.187.)

P

activity. Thus, economy of time,[52] along with the planned distribution of labour-time among the various branches of production, remains the first economic law on the basis of communal production. It becomes law, there, to an even higher degree. However, this is essentially different from a measurement of exchange-values (labour or products) by labour-time.'[53]

With this we come to a question which has been raised very often; that of the operation of the law of value under socialism. It is (or was then) generally known that value counted for the founders of marxism as an economic category which 'expressed the most extensive servitude of the producers to their own products' (*Anti-Dühring*). It is clear from this that they could not possibly have extended the operation of the law of value to a socialist (or a communist) society. On the contrary : any eternalisation of the concept of value was always opposed as a petit-bourgeois utopia. We read in the *Theories* : 'Where labour is communal, the relations of men in their social production do not manifest themselves as "values" of

---

[52] This 'economy of time' is looked at by Marx from another perspective : 'Real economy . . . consists in saving labour-time . . . but this saving identical with development of the productive forces. Hence in no way abstinence from consumption, but rather the development of power, of capabilities of production, and hence both of the capabilities as well as the means of consumption. The capability to consume is a condition of consumption, hence its primary means, and this capability is the development of an individual potential, a force of production. The saving of labour-time is equal to an increase of free time, i.e. time for the full development of the individual, which in turn reacts back upon the productive power of labour itself as itself the greatest productive power. From the standpoint of the direct production process it can be regarded as the production of fixed capital, this fixed capital being man himself.' (*Grundrisse*, pp.711-12.)

[53] *ibid.* pp.172-73. The passage in *Capital* III which is directed against Storch should be understood in just this sense : 'Secondly, after the abolition of the capitalist mode of production, but still retaining social production, the determination of value continues to prevail in the sense that the regulation of labour-time and the distribution of social labour among the various production groups, ultimately the book-keeping encompassing all this, become more essential than ever' (p.851). This, incidentally, is probably the only passage in Marx which such economists as Joan Robinson and Leontiev can legitimately call on in order to impute to him the idea of the 'law of value under socialism'. It is clearly sufficient for them that the phrase 'determination of value' occurs in the sentence. But they could, with equal justification, pick out the scattered passages where Marx, 'to talk the language of Vulgar Economy', speaks of 'capital' in the ancient world (or even under socialism) and claim that for Marx capital is not a historical, but an eternal category.

(Cf. the following passage in *Theories* III, p.257 : 'Labour-time, even if exchange-value is eliminated, always remains the creative substance of wealth and the measure of the cost of its production.')

"things".[54] 'The very necessity of first transforming individual products or activities into *exchange-value*, into *money*, proves two things : 1) That individuals now produce only for society and in society; 2) that production is not *directly* social, is not the "offspring of association", which distributes labour internally.'[55] Thus in a community-producing society 'labour is *posited* as general only through *exchange* ... the mediation' between the private labours of individuals 'takes place through the exchange of commodities, through exchange-value, and through money; all these are expressions of one and the same relation'. By contrast, under socialism, 'the labour of the individual is posited from the outset as social labour . . . He therefore has no particular product. His product is not an *exchange-value*. The product does not first have to be transposed into a particular form in order to attain a general character for the individual. Instead of a division of labour, such as is necessarily created with the exchange of exchange-values, there would take place an organisation of labour whose consequence would be the participation of the individual in communal consumption.'[56] Consequently, the measurement of labour by labour-time would only be a means for social planning[57] (regardless of how important it might otherwise appear for a socialist society) and would naturally have nothing in common with the 'famed value' (Engels) and the law of value.

From what has been said it is clear that the measurement of work by labour-time can fulfil two distinct functions in a socialist society. Firstly, in the production process itself it would serve to determine the amount of living labour required for the production of the various goods, so that labour could be used more economically; and secondly, this measurement can also be regarded as a means of distribution, by which individual producers would be allotted their shares in the social product destined for consumption.

It can, but it does not have to be regarded in this way. For whether the future socialist society resorts to this method of distribution will clearly depend on the degree of development of the social forces of production, i.e. primarily, on 'how much there is to divide

---

54 *Theories* III, p.129.

55 *Grundrisse*, p.158.

56 *ibid.* pp.171-72.

57 Engels comments: 'As long ago as 1844 I stated that the above-mentioned balancing of useful effect and expenditure of labour on making decisions concerning production was all that would be left, in a communist society, of the politico-economic concept of value.' (In *Outlines of a Critique of Political Economy*) 'The scientific justification for this statement, however, as can be seen, was made possible by Marx's *Capital*.' (*Anti-Dühring*, p.368.)

up'.[58] We read in *Capital* : 'The way this division is made will vary with the particular kind of social organisation of production and the corresponding level of social development attained by the producers. We shall assume, but only for the sake of a parallel with the production of commodities', adds Marx, 'that the share of each individual producer in the means of subsistence is determined by his *labour-time*'.[59]

It is clear that what Marx has in mind in this latter instance is a socialist society, 'not as it has *developed* on its own foundations, but on the contrary, just as it *emerges* from capitalist society; which is thus in every respect, economically, morally, intellectually, still stamped with the birth-marks of the old society from whose womb it emerges'. This society has indeed expropriated the capitalists and transformed the means of production into communal property, the people's property; however, it is not yet capable of realising the communist principle of distribution : 'From each according to his ability, to each according to his needs!' Its mode of distribution is therefore still dominated by '*bourgeois right*', which 'is a right of inequality, in its content, like every right'.[60] Therefore 'the individual producer receives back from society – after the deductions[61] have

---

[58] Cf. Engels's letter to C.Schmidt of 5 August 1890: 'There has also been a discussion in the *Volkstribüne* about the distribution of products in future society, whether this will take place according to the amount of work done or otherwise. The question has been approached very "materialistically" in opposition to certain idealistic phraseology about justice. But strangely enough it has not struck anyone that, after all, the method of distribution essentially depends on how much there is to distribute, and that this must surely change with the progress of production and social organisation, and that therefore the method of distribution will also change. But everyone who took part in the discussion described "socialist society" not as something continuously changing and advancing but as something stable and fixed once and for all, which must therefore also have a method of distribution fixed once and for all. All one can reasonably do, however, is 1) to try and discover the method of distribution to be used at the beginning, and 2) to try and find the general tendency of the further development. But about this I do not find a single word in the whole debate !' (*Selected Correspondence*, p.393.)

[59] *Capital* I, p.172 (78-79).

[60] See the significant commentaries on this in Lenin, *The State and Revolution* and Trotsky, *The Revolution Betrayed*, pp.52-54.

[61] Marx lists these deductions as: 'First, cover for replacement of the means of production used up. Secondly, additional portion for expansion of production. Thirdly, reserve or insurance funds to provide against accidents, dislocations caused by natural calamities etc.' Only the remainder of the product therefore is 'intended to serve as means of consumption'. But 'before this is divided among the individuals, there has to be deducted from it : First, the general costs of administration not belonging to production . . . Second,

been made – exactly what he gives to it. What he has given to it is his individual quantum of labour . . . He receives a certificate from society that he has furnished so much labour . . . and with this certificate he draws from the social stock of means of consumption as much as costs the same amount of labour.'[62] In other words, he receives mere certificates of labour the sole purpose of which is to regulate social distribution according to the labour principle. However, there is no room for a law of value in such a society because the form of production which exists here is completely different from commodity production and because the regulation of production and distribution is not left to the blind play of the market, but is subject to the conscious control of society.

It would of course be tempting to go into the question of the operation of the law of value in the Soviet Union and the so-called People's Democracies in this connection. However, this subject goes beyond the framework of our immediate task. In addition we do not think that we could say anything on this question which could measure up to the clarity and depth of the work of E. Preobrazhensky,[63] the most famous economist of the Russian revolution. The main line of his argument consists in the view that any anti-capitalist transformation in an industrially backward country must take place under the conditions of a continuous struggle between the law of value, inherited from the capitalist past, and the diametrically opposed principle of socialist planning; and that the destiny of socialism depends on the outcome of this struggle. And if today numerous economists in the Soviet bloc elevate the law of value to the ranks of a socialist principle of distribution, this shows not only the extent of the theoretical gulf between them and Preobrazhensky and his contemporaries but also how far social and economic relations in the Soviet Union have become separated from the original aims of the October Revolution of 1917.

Let us summarise : the chief distinction between Marx's conception of socialism and that of his predecessors is its scientific character, the fact that he uses the scientific understanding of the present social

that which is intended for the common satisfaction of needs, such as schools, health services etc. Third, funds for those unable to work . . . Only now do we come to the "distribution" . . . namely to that part of the means of consumption which is divided among the individual producers of the co-operative society.' (*Critique of the Gotha Programme. Selected Works*, pp.318-19.) See also *Capital* III, 847-49, 875-76, 877-79.

  [62] *Selected Works*, p.319.

  [63] See *The New Economics*, Moscow, 1926. English translation: Oxford, 1965.

order as the basis for deriving the future socialist vision through an analysis of the capitalist relations of production. The object of the investigation is the same in both instances; namely, modern capitalist society, except that in the one case the concern is with its present-day form, and in the other, with the society of the future which grows out of it. We can therefore see that the economic interrelations studied by Marx must be understood largely as dialectical laws of development. (In fact they can *only* be grasped as such.) This reveals the real meaning of the much discussed 'historicism' of Marx's critique; it is a method which seeks to examine both the conditions of existence and the historical barriers of capitalism. [64] The socialist consequences[65] which follow from this method, and which are aimed at the overthrow of capitalism, are therefore just as fundamental to Marx's system as a whole as his actual study and critique of the economic categories in themselves.

[64] We directed most of our attention to the *Rough Draft*. This explains why we only occasionally mentioned the numerous statements and arguments about communist society in *Capital, Theories* and *Anti-Dühring*.

[65] 'But within bourgeois society, the society that rests on exchange-value, there arise relations of circulation as well as of production which are so many mines to explode it. A mass of antithetical forms of the social unity, whose antithetical character can never be abolished through quiet metamorphosis.' Consequently the enormous significance of proletarian class struggle and the ideological processes which underlie it: 'The recognition of the products as its own, and the judgement that its separation from the conditions of its realisation is improper – forcibly imposed – is an enormous advance in awareness . . . and as much the knell to its doom as, with the slave's awareness that he cannot be the property of another, with his consciousness of himself as a person, the existence of slavery becomes a merely artificial, vegetative existence, and ceases to be able to prevail as the basis of production.' (*Grundrisse*, pp.157, 463.)

29.
The Reification of the Economic Categories and the 'True Conception of the Process of Social Production'

Marx says : 'As the system of bourgeois economy has developed for us only by degrees, so too its negation, which is its ultimate result.'[1] But this has certainly proved to be a long, hard and difficult journey ! It has not only involved the examination and presentation of the development of capital in its concrete shape, but also the step-by-step deciphering of the mystified forms in which capital appears, and the uncovering of their real content. Regarded in this way, the system of bourgeois economics at the same time represents the history of human 'self-alienation', in the study of which we not only have to reveal the alienated character of the economic categories, but also grasp that the 'inversion of subject and object',[2] which is unique to the capitalist mode of production is both necessary and a product of real causes. This is the problem which the young Marx set himself in the *Economic and Philosophical Manuscripts of 1844*, but which was not to find its solution until the completion of *Capital*.

However, this task could not have been accomplished without the basic preliminary work of the great Classical economists – and Marx himself was the first to recognise this fact. We read in the *Theories* : 'Like all economists worth naming . . . Ricardo empha-sises . . . labour as human activity, even more, as *socially determined* human activity . . . it is precisely through the consistency with which he treats the value of commodities as merely "representing" socially determined labour, that Ricardo differs from the other economists.' However, all the Classical economists 'understand more or less clearly (but Ricardo more clearly than the others) that the exchange-value of *things* is a mere expression, a specific social form, of the productive activity of men, something entirely different from things and their use as things, whether in industrial or in non-industrial consump-tion. For them value is, in fact, a simply and objectively expressed

---

[1] *Grundrisse*, p.712.
[2] *Capital* III, p.45.

relation of the productive activity of man, of the different types of labour to one another.'[3]

And this is emphasised even more decisively in the section of the *Theories* devoted to Richard Jones : It states there : 'Even in Ricardo's works' the theoretical analysis 'goes so far that . . . the *independent material form of wealth disappears* and wealth is shown to be simply the activity of men. Everything which is not the result of human activity, of labour, is nature and, as such, is not social wealth. The phantom of the world of goods fades away and it is seen to be simply a continually disappearing and continually reproduced objectivisation of human labour. All solid material wealth is only transitory materialisation of social labour, crystallisation of the production process whose measure is time, the measure of a movement itself.' With Ricardo, however, even 'the manifold forms in which the various component parts of wealth are distributed amongst different sections of society lose their apparent independence. Interest is merely a part of profit, rent is merely surplus profit. Both are consequently merged in profit, which itself can be reduced to *surplus-value*, that is, to unpaid labour.'[4]

However the reification of the social relations of production reaches its high point in the economic 'Trinity' : 'Capital – profit, land – rent, labour – wages', where the capitalist mode of production appears as an, 'enchanted, perverted topsy-turvy world, in which Monsieur le Capital and Madame la Terre do their ghost-walking as social characters and at the same time directly as mere things.'[5] To this extent the classical economists, especially Ricardo, provided a great service, 'in destroying this false appearance and illusion, this mutual independence and ossification of the various social elements of wealth, this personification of things and reification of production relations, this religion of everyday life'.[6]

---

[3] *Theories* III, p.181.

[4] *ibid.* p.429.

[5] Cf. p.28ff above.

[6] Cf. Marx's assessment of the writing of Ricardo's 'proletarian opponent', Thomas Hodgskin : 'The whole objective world, the "world of commodities", vanishes here as a mere moment, as the merely passing activity, constantly performed anew, of socially producing human beings. Compare this "idealism" with the crude, material fetishism into which the Ricardian theory develops in the writing of . . . McCulloch, where not only the difference between man and animal disappears but even the difference between a living organism and an inanimate object. And then let them say that as against the lofty idealism of bourgeois political economy, the proletarian opposition has been preaching a crude materialism directed exclusively towards the satisfaction of coarse appetites.' (*Theories* III, p.267.)

At the same time Marx stresses that even the best of the classical economists, 'remain more or less in the grip of the world of illusion which their criticism had dissolved, as cannot be otherwise from a bourgeois standpoint, and thus they all fall more or less into inconsistencies, half-truths and unsolved contradictions'.[7] And we would add to this that all these economists lack a clear awareness that in general economics deals with reified categories, and that the inverted way in which the social relationships are presented in capitalist production arises necessarily from the essence of this production. However, if they had possessed this awareness they would not have conducted 'political economy' as such, but rather, as Marx did, a 'Critique of Political Economy' – that is, they would have accomplished something which could only be accomplished from the standpoint of the socialist proletariat.

In other words, Marx was the first to succeed in finally overcoming the forms of thinking of bourgeois economics : and it is due to him that we possess the proof that the more the capitalist mode of production develops, the more the social relations of production confront mankind as external, dominating and alien powers.

This process of alienation corresponds to the progressive reification of the economic categories. Marx says this in one of the sections of Volume III of *Capital*[8] dealing with the 'alienation of surplus-value' : 'We have already pointed out the mystifying character that transforms the social relations, for which the material elements of wealth serve as bearers in production, into properties of these things themselves (commodities) and still more pronouncedly transforms the production relation itself into a thing (money). All forms of society, insofar as they reach the stage of commodity production, take part in this perversion.'[9] (It is therefore no accident that the famous chapter dealing with the 'fetishism of commodities' is to be found in the section dealing with commodity circulation in Volume I.)

Of course, this process of reification goes 'further still' in the

[7] *Capital* III, p.830.

[8] *Capital* III, pp.826-32. Cf. the corresponding passage in *Theories* III, pp.482-88.

[9] Cf. *Capital* III, p. 831 : 'In preceding forms of society this economic mystification arose principally with respect to money and interest-bearing capital. In the nature of things it is excluded, in the first place, where production for use-value, for immediate personal requirements, predominates ; and secondly, where slavery or serfdom form the broad foundation of social production, as in antiquity and during the Middle Ages. Here the domination of the producers by the conditions of production is concealed by the relations of domination and servitude, which appear and are evident as the direct motive power of the process of production.

capitalist mode of production : 'If one considers capital to begin with, in the actual process of production, as a means of extracting surplus labour, then this relationship is still very simple, and the actual connection impresses itself upon the bearers of this process, the capitalists themselves, and remains in their consciousness. The violent struggle over the limits of the working day demonstrates this strikingly.'[10] In fact : 'It is quite simple; if with £100, i.e. the labour of 10 (people), one buys the labour of 20 (people) (that is, commodities in which the labour of 20 is embodied), the value of the product will be £200 and the surplus-value will amount to £100, equal to the unpaid labour of 10 (people). Or, supposing 20 workers worked half a day each for themselves and half for capital – 20 half-days equal 10 whole ones – the result would be the same as if only 10 workers were paid and the others worked for the capitalist gratis. Here, in this embryonic state, the relationship is still very obvious, or rather it cannot be misunderstood. The difficulty is simply to discover how this appropriation of labour without any equivalent arises from the law of commodity exchange – out of the fact that commodities exchange for one another in proportion to the amount of labour-time embodied in them – and, to start with, does not contradict this law.'[11]

'The circulation process obliterates and obscures the connection.' For 'whatever . . . the surplus-value extorted by capital in the actual production process and appearing in commodities, the value and surplus-value contained in the commodities must first be realised (*realisiert*) in the circulation process. And both the restitution of the values advanced in production and, particularly, the surplus-value contained in the commodities seem not merely to be realised in the circulation, but actually to arise from it; an appearance which is especially reinforced by two circumstances : first, the profit made in selling depends on cheating, deceit, inside knowledge, skill and a thousand favourable market opportunities; and then by the circumstance that added here to labour-time is a second determining element – time of circulation. This acts, in fact, only as a negative barrier against the formation of value, but it has the appearance of being as definite a basis as labour itself and of introducing a determining element that is independent of labour and results from the nature of capital.'[12]

[10] *ibid.* p.827.
[11] *Theories* III, pp.481-82.
[12] *Theories* III, p.482; *Capital* III, pp.827-28.

A still higher degree of reification is shown 'by capital in its completed form, as it appears as the unity of the circulation process and the production process'.[13] 'Completed capital', gives rise to new formations, 'in which the vein of internal connections is increasingly lost, the production relations are rendered independent of one another, and the component values become ossified into forms independent of one another'. In the first place, 'surplus-value, in the form of profit, is no longer related back to that portion of capital invested in labour from which it arises but to the total capital. The rate of profit is regulated by laws of its own, which permit, or even require, it to change, while the rate of surplus-value remains unaltered. All this obscures more and more the true nature of surplus-value and thus the actual mechanism of capital. This happens to an even greater extent with the transformation of profit into average profit, and of values into prices of production.'

'A complicated social process intervenes here, the equalisation process of capitals, which divorces the relative average prices of the commodities from their values, as well as the average profits in the various spheres of production . . . from the actual exploitation of labour by the particular capitals. Not only does it appear so, but it is true in fact that the average price of commodities differs from their value, thus from the labour realised in them, and the average profit of a particular capital differs from the surplus-value which this capital has extracted from the workers employed by it. The value of commodities appears directly, solely in the influence of a fluctuating productivity of labour upon the rise and fall of the prices of production, upon their movement and not upon their ultimate limits. Profit seems to be determined only secondarily by direct exploitation of labour, insofar as the latter permits the capitalist to realise a profit deviating from the average profit at the regulating market prices which apparently prevail independent of such exploitation.' 'Indeed the basis itself – the determination of the value of commodities by the labour-time embodied in them – appears to be invalidated as a result of the conversion of values into prices of production.'[14]

This fetishistic semblance is still further consolidated 'by the fact that the same process of the equalisation of capital, which gives profit the form of average profit, separates part of it in the form of *rents* as something independent of it and arising from a different foundation, the land. It is true that rent originally emerges as part of profit, which the farmer pays to the landlord. But since this surplus

---

[13] We thus arrive at the subject matter of Volume III of *Capital*.
[14] *Capital* III, pp.828-29. *Theories* III, pp.483.

profit is not pocketed by the farmer, and the capital he employs does not differ in any way as capital from other capitals' (it is precisely because surplus profit is not derived from capital as such that the farmer pays it to the landlord), the land itself, 'appears to be the source of this part of the value of the commodity (its surplus-value) . . . In this formula, in which rent, a part of surplus-value, *is represented in relation to a particular natural element, independent of human labour*, not only the nature of surplus-value is completely obliterated, because the nature of value itself is obliterated; but, just as the source of rent appears to be land, so now *profit* itself appears to be due to *capital as a particular material element of production.* Land is part of nature, and brings in rent. Capital consists of products and these bring in profit. The fact that one use-value which is produced brings in profit, while another which is not produced brings in rent, are simply two forms in which things *produce value*, and the one form is just as comprehensible and as incomprehensible as the other.'[15]

However : It is only 'the division of profit into profit of enterprise and interest (not to mention the intervention of commercial profit and profit from money-dealing, which are founded upon circulation and appear to arise completely from it, and not from the process of production itself), [which] consummates the individualisation of the form of surplus-value, the ossification of its form as opposed to its substance, its essence. One portion of profit [business profit] . . . separates itself entirely from the relationship of capital as such and appears as arising not out of the function of exploiting wage-labour, but out of the wage-labour of the capitalist himself.[16] In contrast thereto, interest then seems to be independent both of the worker's wage-labour and the capitalist's own labour, and to arise from capital as its own independent source.'[17] Consequently the fetish of capital appears in its most complete, and at the same time 'most insane' form in interest-bearing capital.[18] This sketch on the 'alienation of surplus-value', which we have quoted in detail, does not merely offer an excellent summary of the content of all three volumes of *Capital*. More than this, it shows what the essential result of Marx's *Critique of Political Economy* consists in : namely, in the

---

[15] *Theories* III, pp.483-85.
[16] 'The labour of exploiting is identified here with the labour which is exploited.' (*ibid.* p.495.) Besides, in most instances the 'labour of exploiting' is not carried out by the capitalist himself, but by his manager.
[17] *Capital* III, p.829.
[18] *ibid.* p.465.

proof that economics, *'is not concerned with things, but with relations between people, and in the last instance between classes'*; but these relations 'are always *bound to things* and *appear as things'* (Engels). The epoch-making significance of this discovery is immediately apparent. Only in this way could Marx posit in the place of the reified categories of bourgeois economy, a 'true conception of the process of social production'[19] – in the sense meant in Galiani's fine expression : 'The real wealth is man ... himself.'[20] Only in this way could the science of political economy be turned into a social science. As already stated in the *Rough Draft* : 'When we consider bourgeois society in the long view and as a whole, then the final result of the process of social production always appears as the society itself, i.e. the human being itself in its social relations. Everything that has a fixed form, such as the product etc., appears as merely a moment, a vanishing moment, in this movement. The direct production process itself here appears only as a moment. The conditions and objectifications of the process are themselves equally moments of it, and its only subjects are the individuals, but individuals in mutual relationships, which they equally reproduce and produce anew. [It is] the constant process of their own movement in which they renew themselves even as they renew the world of wealth they create.'[21]

[19] *Grundrisse*, p.711.
[20] See *ibid.* p.846 and *Theories* III, p.267.
[21] *Grundrisse*, p.712.

PART SEVEN
Critical Excursus

30.
The Dispute Surrounding Marx's Schemes of Reproduction

The main aim of this work has been of a methodological nature. We set out from the position that previous research was excessively concerned with the material content of Marx's economic work, and exhibited far too little interest in his specific method of investigation.* We therefore tried to show how much the *Rough Draft* has to teach on the subject of methodology. But if this is true then the methodological insights which can be gained from a study of this work should also throw a new light on certain of the old disputes in marxist economics – in particular, the much-discussed question of the schemes of reproduction in Volume II of *Capital,* and the so-called problem of realisation.

I. INTRODUCTION

1. A note on the formal aspect of the schemes of reproduction in Volume II

In order to facilitate the following representation of this question we want, to begin with, to deal briefly with the form – that is the

---

* This characterised the method of study of the bourgeois economists, who were reproached by Marx for their 'crude concentration on the material'

numerical form – of the schemes of reproduction in Volume II.

In his presentation of the conditions for the reproduction of the total social capital Marx divided social production into two large Departments, of which Department I produces means of production, and Department II articles of consumption. The value of the product in each Department is divided into c + v + s; that is, the constant capital used in one phase of production, the variable capital expended in wages, and finally the surplus-value created in this phase of production. He then examines to what extent the value components of the product of each of the Departments must be mutually exchanged, so that the next phase can take place.

The first question is that of the conditions which make simple reproduction possible (i.e. reproduction on a constant scale). Marx drew up the following scheme to illustrate this :

$$\begin{array}{lll} \text{I} & 4000c + 1000v + 1000s = 6000 \\ \text{II} & 2000c + 500v + 500s = 3000 \end{array}$$

Since under the conditions of simple reproduction, Department I requires exactly as much constant capital as it used in the preceding period of production, i.e. 4000c, it can cover these 4000 units by means of its own production, without having to resort to exchange with Department II.

Similarly, Department II, whose products consist of consumer articles, can directly use 500v and 500s, which it employs for the personal consumption of the workers and capitalists, without exchange with Department I. However, what does have to be exchanged between both Departments is that part of the product of Department II whose value corresponds to its constant capital, and that part of the product of Department I which is equal to its variable capital and its surplus-value. Hence the general formula for equilibrium in the simple reproduction of social capital is clearly :

$$c\ II = v\ I + s\ I,$$

i.e. the constant capital used by Department II must be the same size as the variable capital plus the surplus-value in Department I.

Nevertheless, the above formula cannot be applied to conditions of extended reproduction, i.e. it cannot be applied when a part of the

and their lack of interest 'in understanding the distinctions of form of the economic relations'.

surplus-value accumulates, and instead of being consumed by the capitalists is added to the constant and variable capital of each Department. We can adopt Bukharin's method of denoting that portion of surplus-value to be consumed by the symbol α, the portion of surplus-value to be added to constant capital in the succeeding production period by βc, and the portion to be added to variable capital by βv; in this case the previous formula for equilibrium must be altered as follows in order to correspond to the conditions of extended reproduction.

$$c\,II + \beta\,c\,II = v\,I + \alpha\,I + \beta\,v\,I,^1$$

This is in fact the general formula which forms the basis of Marx's scheme of reproduction in Chapter XXI of Volume II of *Capital*.

We find two schemes in this chapter which, according to Marx, were meant to illustrate the process of accumulation at two different stages of capitalist development.[2] The first is as follows (after rounding off Marx's fractions, and expressing it in Bukharin's notation) :

		c		v		α		βc		βv
Year 1	I	4000	+	1000	+	500	+	400	+	100
	II	1500	+	750	+	600	+	100	+	50
Year 2	I	4400	+	1100	+	550	+	440	+	110
	II	1600	+	800	+	560	+	160	+	80
Year 3	I	4840	+	1210	+	605	+	484	+	121
	II	1760	+	880	+	616	+	176	+	88
Year 4	I	5324	+	1331	+	666	+	532	+	133
	II	1936	+	968	+	677	+	194	+	97

and so on. The second scheme, which corresponds to a more advanced stage of capitalist development, proceeds under the assumption of a higher organic composition of capital. Moreover, in contrast to the first scheme, the composition of capital is the same in both Departments (namely 5c : 1v). In the second scheme reproduction takes place as follows :

---

[1] This formula can be found in Bukharin's *Imperialism and the Accumulation of Capital*, p.158.

[2] *Capital* II, p.514.

	c		v		α		βc		βv
Year 1	5000	+	1000	+	500	+	417	+	83
	1430	+	285	+	101	+	153	+	31
Year 2	5417	+	1083	+	542	+	452	+	90
	1583	+	316	+	158	+	132	+	26
Year 3	5869	+	1173	+	587	+	489	+	98
	1715	+	342	+	171	+	143	+	28
Year 4	6358	+	1271	+	636	+	530	+	106
	1858	+	370	+	185	+	155	+	30

and so forth.

Both schemes were sharply criticised by Rosa Luxemburg. She maintained against the first that Marx could only obtain the 'precise logical rules which specify the relations of accumulation in Department I' at the cost of 'any kind of principle in the construction of these relations in Department II', by allowing this Department to accumulate and consume without any 'visible rule' and merely in 'an erratic fashion'.[3] Luxemburg does concede that accumulation in the second scheme proceeds in an orderly way in both Departments, so that such 'arbitrary changes in the distribution of surplus-value in II' no longer occur there. However, she thought it was possible to maintain that even in this instance, 'accumulation in Department II is completely determined and dominated by accumulation in Department I . . . that Department I has taken the initiative and actively carries out the whole process of accumulation while Department II is merely a passive appendage.'[4]

As far as the second objection is concerned (the validity of which, strangely, has never been challenged by the marxist camp), this has been convincingly refuted by Joan Robinson, who proved that the 'arithmetic is perfectly neutral between the two Departments', and that the impulse to accumulation could come equally well from either Department.[5]

However, accumulation in Marx's first diagram is by no means as 'variable' and 'erratic' as it seemed to Luxemburg, and as has generally been thought up to now. Because, if the first year is disregarded, it can be seen that in this scheme Department I constantly

[3] Luxemburg, *The Accumulation of Capital*, p.122.
[4] *ibid.* p.127.
[5] Joan Robinson's introduction to the English edition of *The Accumulation of Capital*, p.19.

accumulates 50% of the surplus-value and Department II 30%. This is, of course, no accident, but rather the necessary result of the difference in the organic composition of capital in each of the Departments. It can be algebraically shown[6] that − if the rate of surplus-value is the same in both Departments and does not change in the course of reproduction − the formula for equilibrium

$$c\ II + \beta c\ II = v\ I + \alpha\ I + \beta\ v\ I$$

requires a strict correlation between the rate of accumulation and the organic composition of capital in each Department. If we assume, with Marx, that the organic composition of capital and the rate of accumulation remain the same in succeeding periods of production, then the rates of accumulation in both Departments must move in inverse proportion to the rates of organic composition, or, expressed in an equation :

$$\frac{\beta\ I}{s\ I} \quad : \quad \frac{\beta\ II}{s\ II} \quad = \quad \frac{v\ II}{c\ II + v\ II} \quad : \quad \frac{v\ I}{c\ I + v\ I}$$

Thus in Marx's first diagram the relation between the rates of accumulation in both departments was 50% (Department I) : 30% (Department II). The relation of v :c was $\frac{1}{3}$ in Department II and $\frac{1}{5}$ in Department I. Since $5 : 3 = \frac{1}{3} : \frac{1}{5}$, the required conditions for equilibrium of reproduction are given.

However, this is enough on Rosa Luxemburg's 'mathematical' error. She was probably distracted by the form of Marx's numerical example, which in fact does appear to be rather difficult and confusing. The confusing aspect consists in the fact that in both schemes accumulation in the initial year does not follow the rule which guides accumulation in the succeeding years. We can only guess at why Marx chose this form of presentation; perhaps it was simply a question of a preliminary attempt, which he found no time to correct.

Strangely enough, the cumbersome form of Marx's diagrams also misled Luxemburg's most severe critic, Bukharin. As already mentioned, it was Bukharin who first formulated the general relation of equilibrium for extended reproduction $c\ II + \beta\ c\ II = v\ I + \alpha\ I + \beta\ v\ I$. However, he derived two other, totally incorrect formulae from

---

[6] I have to thank my friend, the statistician H. Chester of Detroit, for the mathematical proof of this relation.

this one, namely : c II $= $ v I $+ \alpha$ I, and β v I $= \beta$ c II.[7] It is in fact correct that in the *initial year* of Marx's first scheme c II $=$ v I $+ \alpha$ I and also β v I $= \beta$ c II. However, this is only the case because Marx was not able to ascertain directly the correct proportion between c I and c II. In all the succeeding years of the first scheme, and in all the years of the second, c II is necessarily smaller than v I $+ \alpha$ I, and β c II greater than β v I. In other words, Bukharin completely forgot that the extended reproduction of the total social capital must not only lead to the growth of c and v but also to that of α, i.e. to the growth of the individual consumption of the capitalists. Nevertheless, this elementary mistake remained unobserved for almost two decades[8] and Bukharin was generally regarded as the most authoritative defender of marxist 'orthodoxy' against Rosa Luxemburg's attacks 'on those parts of Marx's analysis, in which the incomparable master has handed down to us the completed product of his genius'.[9] Nevertheless, Bukharin's general formula for equilibrium is very useful, although he too (like most critics of Rosa Luxemburg) mistook the mere formulation of the problem for its solution.

2. The 'approximation to reality' of Marx's schemes of reproduction

This is sufficient on the form of Marx's schemes of reproduction. When we come to examine their content we must first be clear on the question as to whether and to what extent Marx wanted to use these schemes to describe processes taking place in the real world of capitalism.

Curiously, only very few marxists have tried to deal with this question. If, for example, one follows the discussions on the schemes of reproduction which revolved around Luxemburg's book then one immediately comes across a strange paradox. The Austro-marxists, the opponents of Luxemburg (i.e. Kautsky, Bauer, Eckstein,

[7] Bukharin, *op. cit.* p.158.

[8] It was first noted by Sweezy in his *Theory of Capitalist Development*, 1942, p.164.

[9] Bukharin, *op. cit.* The present-day reader may find aggressive and often frivolous Bukharin's tone somewhat unpleasant, when one remembers that Rosa Luxemburg had fallen victim to fascist murderers only a few years previously. That his tone was dictated more by political than scientific interests provides some explanation. Bukharin saw his task as that of breaking the still very strong influence of 'Luxemburgism' within the German Communist Party (KPD), and any means seemed justified.

Hilferding and others), knew, of course, only too well that Marx's schemes were conceived at the highest level of abstraction and therefore ignored many key features of capitalist reality – such as the existence of non-capitalist classes, and areas of the world, external trade, the average rate of profit, prices of production which diverge from values, etc . . . Yet despite this, all these authors looked for concrete proof of the unlimited economic viability of the capitalist form of economy in these very same schemes!

Let us begin with the founder of the Austro-marxist school, Karl Kautsky. In his magnum opus he strongly criticises Rosa Luxemburg's 'hypothesis' that capitalism must break down for economic reasons; he asserts that Luxemburg 'finds herself in opposition to Marx, who proved the opposite in the second volume of *Capital*, i.e. in the schemes of reproduction'.[10]

Kautsky did not in fact come to this conclusion until after the First World War. However, members of his school proposed similar views much earlier. As we shall see, Rudolf Hilferding's interpretation of Marx's schemes in 1909 (in *Finanzkapital*) amounted to saying that according to these schemes, capitalist production – given the correct proportions between the individual branches of production – 'could be extended indefinitely . . . without leading to an overproduction of commodities'.[11] And at the Vienna Conference of the *Verein für Sozialpolitik* in 1926, Hilferding reminded his academic audience that he had always been an opponent of the 'breakdown theory'. He declared, 'I consider that on this point I find myself in complete agreement with the theories of Karl Marx, to whom a breakdown theory is always falsely attributed. The second volume of *Capital* shows how production is possible on an ever-extended scale inside the capitalist system.' He then added jokingly : 'I've often thought that it is not at all so bad that this second volume is so little read, since, under certain conditions, it could be interpreted as a hymn of praise to capitalism.'[12]

We can see that Hilferding, too, wishes to derive a direct refutation of the 'breakdown theory' from the reproduction schemes in Volume II; he too confuses a numerical illustration with a theoretical proof, and on top of that he confuses the sphere of the 'abstract' with that of the 'concrete'! Otto Bauer proceeds in much the same way.

[10] Karl Kautsky, *Die Materialistische Geschichtsauffassung* Vol.II, pp.546-47.
[11] Cf. p.483ff of this chapter.
[12] Taken from Grossmann's *Akkumulations- und Zusammenbruchsgesetz des kapitalistischen Systems*, pp.57-58.

Although he concedes to Luxemburg that 'the figures which Marx uses in Volume II of *Capital* for the presentation of the reproduction process . . . are arbitrarily chosen and not free from contradictions' he goes on to say : 'However, the fact that Marx's presentation is open to objection does not mean that this line of reasoning is in itself false.' With this in mind Bauer drafted his own 'non-arbitrary' scheme of reproduction, and continued to assert that the figures which he constructed 'demonstrate' or 'prove' his interpretation of Marx's theory of reproduction, interpreted as meaning the capacity of the capitalist mode of production to extend itself without limit.[13]

The weakest of Luxemburg's critics, Eckstein, even manages to lump together two perspectives – the purely theoretical and the empirical – in one and the same sentence. Thus we read at the beginning of his discussion : 'If one wishes to study the problem of crises,[14] one must above all pose the question of how the reality of capitalist accumulation relates to Marx's schemes for equilibrium, which only demonstrate the *possibility* of equilibrium.' But following this, on the next page, he writes : 'Marx's schemes show how capitalist production must proceed if it is to remain in equilibrium – they show, in actual fact, the size of the social requirement for the various types of product.' And on the following page to this : 'The capitalist mode of production is guided by the search for profit. The question is, then, whether Marx's schemes show how this profit is realised for the capitalists. This is the case throughout . . . The schemes show exactly who buys the products.'[15]

The foregoing examples are sufficient to illustrate our point. They show just how right Henryk Grossmann was when he wrote : 'The neo-Harmonists glorify the schemes for equilibrium, not because they are a particularly excellent methodological tool of analysis, but rather because they [i.e. the neo-Harmonists] – in confusing the method of analysis with the phenomena to be analysed – considered themselves to have found within the schemes a real tendency to equilibrium in capitalism.'[16]

How can we explain this mistake on the part of the Austro-marxists? How could they be guilty of such an elementary error?

It would be too easy to say that the 'wish was father to the

[13] See Otto Bauer, *'Die Akkumulation des Kapitals'*, in *Die Neue Zeit*, 1913, pp.836, 866.

[14] As we can see, Eckstein confused the problem of the reproduction of social capital with that of crises.

[15] This is discussed in Eckstein's review, printed in the 'Appendix' to Luxemburg's book in the 1923 German edition, pp.487, 488, 489.

[16] Grossmann, *op. cit.* p.95.

thought', and that the Austro-marxists, who were totally immersed in reformist practice, instinctively resisted the idea of an economic breakdown of the prevailing social order (in the same way that they could not grasp the historical necessity of the breakdown of the Austro-Hungarian Empire and the defeat of the Central Powers). This unconscious motive certainly played a prominent role. We believe, however, that their error must also be attributed to a lack of understanding of Marx's economic methodology.

If this methodology is regarded as a positivist science (i.e. if it is divested of its dialectical character), then it is clearly difficult to distinguish Marx's economic method from the conceptual procedure of 'academic theory', which first of all eliminates individual and particular features of economic phenomena (method of 'abstraction') in order to re-introduce these same features in successive stages (method of 'successive concretisation' or 'approximation'). However, since these individual and particular features are only eliminated and reintroduced 'externally', i.e. without any kind of dialectical mediation, the illusion can easily arise that there is no qualitative 'bridge' between the 'abstract' and the 'concrete'.[17] One could well adopt the view that the theoretical model in fact contains all the essential elements of the concrete object of study (although in simplified form) – in much the same way as an aerial photograph shows all the fundamental elements of scenery, although only ranges of mountains, large rivers and forests etc. are visible on it. If the reciprocal relation between 'abstract' and 'concrete' is understood in this way, then the necessary 'contradiction between the general law and further developments in the concrete circumstances'[18] must be overlooked; one falls into the illusion that the abstract picture simply reflects the concrete circumstances, without any form of 'mediation'. And in our opinion this was precisely the methodological source of the error made by Luxemburg's Austro-marxist critics. Namely, they forgot that the abstract formulae in Volume II of *Capital* simply represent 'one stage of the analysis',[19] cannot therefore be directly applied to concrete capitalist reality, and first require numerous 'intermediary links'. In other words : The Austro-marxists mistakenly compounded

---

[17] 'The procedure of common-sense finite cognition here is that it takes up again equally externally from the concrete that which it had left out in the abstractive creation of this universal. The absolute [dialectical] method on the other hand does not hold the position of external reflection; it draws the determinate element directly from its object itself, since it is the object's immanent principle and soul.' (Hegel, *Science of Logic* Vol.II, p.472.)

[18] *Theories* III, p.87.

[19] Cf. Trotsky's view, cited on p.428 above.

two different phases of Marx's analysis, and inevitably ended up on the wrong track. Their disregard of Marx's dialectics thus took its eventual revenge on them !

3. The basic presupposition of Marx's schemes of reproduction

We have already pointed out in Chapter 3 that the category of use-value also enters into the social relations of reproduction. Marx says on this, right at the beginning of his analysis of the reproduction process in Volume II : 'So long as we looked upon the production of commodities and the value of the product of capital individually, the bodily form of the commodities produced was wholly immaterial for the analysis, whether it was machines, for instance, corn or looking-glasses. It was always but a matter of illustration, and any branch of production could have served that purpose equally well . . . So far as the reproduction of capital was concerned, it was sufficient to assume that that portion of the product in commodities which represents capital-value finds an opportunity in the sphere of circulation to reconvert itself into its elements of production and thus into its form of productive capital; just as it sufficed to assume that both the worker and the capitalist find in the market those commodities on which they spend their wages and the surplus-value.' However : 'This merely formal manner of presentation is no longer adequate in the study of the total social capital and of the value of its products. The reconversion of one portion of the value of the product into capital and the passing of another portion into the individual consumption of the capitalist as well as the working class form a movement within the value of the product itself in which the result of the aggregate capital finds expression; and this movement is not only a replacement of value, but also a replacement in material and is therefore as much bound up with the relative proportions of the value-components of the total social product as with their use-value, their material shape.'[20]

Our reason for quoting this passage at such length, although it was not developed any further in the final version of Volume II, is that we consider that it provides a guide to a better understanding of Marx's schemes of reproduction.[21] What Marx had in mind here was evidently the antithesis between use-value and exchange-value,

[20] *Capital* II, p.398.
[21] A section from 'Manuscript VIII' follows this section from 'Manuscript II' in the text of Volume II which was edited by Engels.

which we have often mentioned in the course of this work, which we have already encountered in the analysis of value and money, and which in fact penetrates the entire system of bourgeois economics. Admittedly, Marx's examination of the process of production and circulation of individual capital could give rise to the impression that the sole objective of capitalist production is the creation of value and surplus-value. Now however, in the study of the reproduction of social capital, it turns out that this creation of value and surplus-value collides with a barrier which was not taken into consideration in the foregoing analysis – the barrier of 'use-value on a social scale'.[22] In order to reproduce its capital, the 'society', i.e. the 'total capitalist', must not only have a fund of values at its disposal, but also find these values available in a particular useful form – in the form of machines, raw materials, means of subsistence – and all this in the proportions determined by the technical requirements of production. Thus the creation of value and surplus-value is already technically bound up with the 'social change of matter' (*Stoffwechsel*), even disregarding the necessity of disposing of the commodities produced, that is finding buyers for them.

But does this actually mean that in the final analysis the capitalist economic order does after all have as its aim the satisfaction of society's production and consumption requirements? Not at all. The most conspicuous feature of this economic order is, and remains, its insatiable urge for constantly growing profits. Consequently, only those 'goods' are produced as are at the same time values; and so material human needs are only satisfied to the extent that satisfying them appears indispensable for increasing surplus-value. Thus, for example, the creators of all social wealth, the workers, have large (and fortunately constantly growing) needs; however, they can only satisfy these needs if their labour-power is a saleable commodity on the market, and it can only be sold if it proves itself to be capable of creating surplus-value. The same applies to the so-called 'objective factors' : even the most perfect machines and techniques of production are only employed if they promise to raise the rate of profit. And finally even the 'total capitalist' himself is limited in his comforts and pleasures by the necessity for constant accumulation. Thus, even if the category of value seems conditioned by that of use-value when regarded from the standpoint of the social process of reproduction, this latter process is totally subject to value and the creation of value in the capitalist economy. And it is this very antinomy between con-

[22] *Capital* III, p.636.

tradictory aims, the constant interaction of the categories of value and use-value – which must however, be reconciled ! – which should be kept in mind when we are discussing the reproduction of total social capital, as analysed by Marx.

Of course the fundamental possibility of a solution to this antinomy can only be demonstrated by a highly abstract, yet simple, model : Marx's schemes of reproduction provide just such a model in that they divide total social production into two large departments – means of production and means of consumption – and set each department to work for the other. In order to be able to repeat the production process, each of the two departments has primarily to attend to the replacement of the value of its elements of production; however, it can only do this if it obtains a portion of these elements of production from the other department, in a suitable material form. However, on the other hand, each department can only come into the possession of the use-values which it requires if it obtains them from the other department by means of the exchange of value-equivalents. This mutual dependence between society's 'replacement of value' and its 'replacement of material,' is clearly expressed in the schemes of reproduction; but the schemes can only show this dependence by strictly separating the two departments from one another and confining their mutual relations exclusively to the exchange of value-equivalents. Thus the alleged 'rigidity' of the schemes' basic presuppositions corresponds exactly to the problem which they first had to solve; and if numerous theoreticians (Tugan-Baranovsky, Otto Bauer and others) attempt to 'improve' Marx's schemes by the introduction of less strict conditions to make them more real, this merely proves how little they have grasped the meaning and the structure of these schemes.

Admittedly, one could question the point of presenting the possibility of a solution to the conflict between use-value and value, as revealed in the process of social reproduction, when this solution occurs millions of times over in capitalist practice by the adjustment of the prices of commodities to social demand, and the bankruptcy of individual firms. Certainly actual capitalist practice shows us the phenomena of economic crises, in which the periodically recurring impossibility of a solution to the conflict is demonstrated, and in which 'the contradictions and antagonisms of bourgeois production are strikingly revealed'.[23] Looked at in this way the question as to what extent the antinomy between use-value and value can be over-

---

[23] *Theories* II, p.500.

come at all within the capitalist economic order is certainly of theoretical interest, and the schemes of reproduction in Volume II which serve as an answer to this question, can be of great service in this respect.

4. The schemes of reproduction and the realisation problem

So much on the basic presupposition of the schemes of reproduction in Volume II, namely, that the relations of exchange between the two main departments of social production must accord with one another, both from the point of view of value, and from that of use-value, if equilibrium conditions for the reproduction of total social capital are to be maintained. (We have to stress this necessary condition of the schemes, since it is, unfortunately, too often forgotten in marxist literature.)

Naturally, this is not the only aspect which a study of the schemes offers and not the only task which Marx set himself by constructing them! His main aim – following Quesnay's example – was to design a new *Tableau Economique*, which was to summarise the 'innumerable individual acts of circulation', on the surface of bourgeois society, 'in their characteristic social mass movement', that is, 'the circulation between the great functionally determined economic classes of society'.[24] Marx's comments on Quesnay's *Tableau* also, therefore apply to the schemes of reproduction in Volume II of *Capital*. They too have the aim 'of portraying the whole production process of capital as a process of reproduction' (in which circulation appears as a mere form of the reproduction process), and at the same time of including in this reproduction process not only the 'origin of revenue and the exchange between capital and revenue', but also the 'relation between reproductive consumption and final consumption' and 'the circulation between consumers and producers'.[25] The only difference is that Marx's scientific project was incomparably more complex and difficult than Quesnay's! In the first place, as far as Quesnay was concerned, value still coincided with use-value,[26] and so the funda-

---

[24] *Capital* II, p.363.

[25] *Theories* I, pp.343, 344.

[26] The Physiocrats' 'method of exposition is, of course, necessarily governed by their general view of the nature of value, which to them is not a definite social mode of existence of human activity (labour), but consists of material things – land, nature, and the various modifications of these material things.' (*ibid.* p.46.)

mental question of a contradiction between use-value and exchange-value simply did not exist for him. And secondly, Quesnay dealt merely with simple reproduction, whereas Marx's central concern was, necessarily, the extended reproduction of total social capital. Accordingly Marx's schemes of reproduction are not only supposed to demonstrate how all the component parts of the annual value-product of society (c + v + s) mutually replace each other, but also how a portion of the surplus-value produced can be devoted to the further expansion of capitalist production – which naturally pre-supposes the regular exchange of these value-components, and their realisation on the market. In this sense the schemes of reproduction in Volume II can be regarded as a (provisional) solution to the so-called problem of realisation.

During the course of its history political economy has offered three solutions to the problems of realisation.

The first solution goes back to J. Mill, D. Ricardo and J. B. Say. All these economists thought they could solve the problem of the realisation of surplus-value by equating capitalist production with simple commodity production, but naïvely reduced the latter to the simple exchange of products. Since any act of production – they teach – creates its own demand, and since in the last analysis products are always exchanged for products, there is a 'metaphysical equilibrium' of sellers and buyers. So in the last analysis all commodities can be disposed of on the market – provided they are produced in the right quantities, in correct proportions. For these economists the realisation problem as such did not really exist, and was reducible in fact to the problem of the proportionality between the different branches of social production.

The position of a contemporary critic of the classical school, Sismondi, was quite different. Sismondi was the first bourgeois economist to be aware of the historically specific character of the capitalist type of economy. Thus he regarded the commodities appearing on the market not simply as 'products' but as products of capital. That is : the owner of the capital obtains an increase in value (*mieux valeur*) in their production, not 'because the product of his enterprise yields more than the amount of production costs, but because he does not pay the full production costs, because he gives the worker an insufficient wage for his labour'.[27] It is precisely this increase in value, this 'surplus-product' which forms the source of the accumulation of capital. But how can the surplus-product be sold

[27] See Sismondi, *Nouveaux Principes de l'Economie Politique*, Vol.I, Book 2, Chapter 4 ('How capitalist profit arises'), p.92.

if the workers who have produced it can only buy back that part of the product which corresponds to the wage for their labour, and if the capitalists themselves do not consume the entire surplus-product, since a portion of it must be capitalised? Sismondi regarded this as an insuperable difficulty; he thought that in the final analysis the realisation of the surplus-product would be impossible – unless it were disposed of and thus realised, abroad.

What was Marx's solution to the problem? It can be regarded as a unique synthesis of the views of Ricardo and Sismondi. Marx in no way denied that the realisation of surplus-value represented one of the thorniest problems of bourgeois economics. However, he categorically rejected Sismondi's doubts as to the possibility of realisation. According to Marx, capitalist production does in fact create its own market and in this sense also 'solves' the problem of the realisation of surplus-value. It does not solve this problem by completely abolishing it, but by 'creating the form', in 'which it can move.' i.e. by 'transferring . . . the difficulties of realisation . . . to a wider sphere . . . giving them greater latitude'.[28] ('This is', we read in *Capital*, 'in general the way in which real contradictions are resolved').[29] The dialectical solution of the realisation problem can, therefore, only lie in the advance of the capitalist mode of production, in the constant extension of its internal and external market. Looked at from this perspective the extended reproduction of capital is neither 'impossible' (as it seemed to Sismondi), nor can it proceed (as the Classical economists thought) *ad infinitum*, since the capitalist mode of production must reproduce its internal contradictions at a continually higher level, until the 'spiral' of capitalist development (an image borrowed from Sismondi) reaches its end.

This dialectic of the realisation problem should be kept in mind if we really want to appreciate the breadth and significance of half a century of debate on the schemes of reproduction in Volume II of *Capital*.

[28] *Capital* II, p.473.
[29] *Capital* I, p.198 (104). Cf. *Capital* III, p.250 : 'Capitalist production seeks continually to overcome these immanent barriers, but overcomes them only by means which again place these barriers in its way and on a more formidable scale' – this refers to the falling rate of profit and the devaluation of capital.
This is the sense in which Marx employed the concept of 'living contradiction', which he took from Hegel. See *Grundrisse*, p.421, 774-75. In addition, Marx's letter to Johann Schweitzer of 24 January 1865, in *Selected Correspondence*, pp.142-49.

II. THE DISCUSSION BETWEEN THE 'NARODNIKS' AND THE 'LEGAL' MARXISTS IN RUSSIA

Curiously, the schemes of reproduction in Volume II of *Capital* remained unnoticed by German marxist literature for almost two decades. Kautsky, the sole writer who did deal with this subject, devoted a mere two lines to them in a review of Volume II of *Capital* (1885) : 'Finally, the accumulation of surplus-value, the expansion of the production process brings further complications.'[30] And this was literally everything which was said on the schemes of reproduction during those decades. It was not until a translation of Tugan-Baranovsky's book was published in Germany that the attention of marxist theoreticians there was drawn to Marx's analysis of the process of social reproduction : it was first extensively discussed in Hilferding's *Finanzkapital* (1909).

This strange state of affairs is perhaps not too difficult to explain. There was clearly no particular contemporary social reason arising from conditions in West and Central Europe which would have prompted the theoreticians of the Second International to discuss the contents of Part III of Volume II of *Capital*. And so this volume simply remained in total oblivion.

The situation was completely different in Russia, where the date of the publication of Volume II of *Capital* coincided with the period of debate on the subject of the possibility and necessity of capitalist development in Russia, a debate which particularly occupied the minds of the progressive intelligentsia. Both camps in the dispute appropriated the analyses in Volume II – the Narodniks who denied the possibility and the marxists who insisted on it – in order to try and find the means to resolve the question which was of such vital importance to them.[31] And so Russia became the first country in which the crucial theoretical significance of these analyses emerged.

1. Engels's debate with Danielson

The most famous of the Narodnik theoreticians was the translator of *Capital*, N. Danielson, who engaged in a lively correspondence with Engels after Marx's death.

[30] Quoted by Rosa Luxemburg.
[31] We can only deal cursorily with this controversy here; we would refer the reader to the brilliant presentation in Luxemburg's *The Accumulation of Capital*.

In his letter of 3 February 1887 Danielson informed Engels that he intended to write a book, 'which would offer the reading public an explanation of our economic life and its developmental tendencies in the light of the "Author's" [i.e. Marx's] theories'.[32] Engels naturally agreed to Danielson's scheme, and stressed how important it would be, 'to show how our author's theory could be applied to your conditions'.[33] However four years had elapsed, with Danielson's work on the book under way, before he and Engels began a vigorous discussion which immediately revealed the extent of the differences in their theoretical assumptions.

Danielson wrote on 24 November 1891 : 'In my last letter I wanted to show you a Russian version of the "creation of the internal market for the industrial classes". I wanted to demonstrate how the destruction of the subsidiary trades of the countryside, the process whereby manufacture is divorced from agriculture, takes place, in order to prove that it is "only the destruction of rural domestic industry which can give the internal market of a country that extent and stability which the capitalist mode of production requires".[34] I wanted to direct your attention to our unique situation. We make our appearance on the world market at a time when the capitalist mode of production and the technical progress which it brings with it, have won the upper hand . . . As a result, we have, on the one hand, a peasantry which is growing ever poorer, and on the other hand, an industrial sector which is becoming continually more concentrated and technically advanced, but which is completely dependent on the fluctuations of the internal market, i.e. dependent on the extent of the separation of industry and agriculture.'[35]

Thus, Danielson's doubts as to the possibility of a full development of capitalism in Russia are already evident in this letter. But his scepticism on this subject is first really revealed in a letter of 24 March 1892. He writes : 'We "liberated" between 20 to 25 per cent of our rural population from the land. Those peasants now wander around looking for work . . . What can they do? Go into factories? But we know that the number of workers employed in present-day industry is constantly declining[36] . . . How many workers can our own internal *market* absorb, until it is totally satiated? "As buyers of com-

[32] *Perepiska K.Marksa i. Fr.Engelsa s russkimi politisheskimi dejatelami*, (*Marx and Engels's Correspondence with Russian Political Figures*), 1947 p.106.

[33] *ibid.* p.107.

[34] Quoted from *Capital* I, p.911 (745).

[35] *Perepiska*, pp.119-20.

[36] Danielson forgets that Marx referred to a relative, not an absolute decline of the productively employed workers.

modities the workers are important for the market. But as sellers of their commodity – labour-power – capitalist society tends to keep them down to the minimum price"[37] . . . here we have our starting point – our internal market. . . . A capitalist nation resolves the contradiction pointed out by our author by the extension of its external markets. But how can we escape this contradiction? Just as it is impossible to imagine a capitalist factory whose production were to be dedicated exclusively to the consumption of the workers employed by it, so a capitalist nation without foreign markets appears equally impossible.[38] And it is precisely for this reason that every capitalist country summons all its powers to conquer the markets of its rivals; there is no capitalism without markets.' But could Russia obtain external markets? 'We enter the world arena at a time when all the efforts of our competitors are stretched to the utmost, when they have to be content with even the slightest rate of profit . . . It therefore seems that our beloved child – capitalism, which destroys the basis of peasant domestic industry, but which has neither an internal, nor an external market, possesses no solid foundation for its development.'[39]

It is clear that Danielson is wrong here; but what is the source of his error? He was correct when he asserted that neither the workers of an individual capitalist factory, nor those of a capitalist country would be in a position to 'buy back' the product of their labour; rather, that they can always buy only the part of this product which corresponds to their wage (not v + s, but only v); and he was also correct when he saw the disproportion between the total sum of wages and the size of the value-product newly created by the workers, as constituting one of the glaring contradictions of the capitalist mode of production.[40] However, one should not regard this question statically, as he did! For, as long as accumulation progresses, and a portion of accumulated surplus-value is used to employ additional labour-power, i.e. workers, then these will help to realise the surplus-value created in the previous period of production by spending their wages. Of course, the newly employed workers will also in turn create a value-product, whose size must exceed the total sum of their wages,

[37] Quote from *Capital* II, p.320.

[38] Danielson expresses the same idea in the later published *Outlines* as follows: 'Just as an individual factory owner could not maintain himself as a capitalist even for a day if his market were confined to his own requirements and those of his workers, so the home market of a developed capitalist nation must also be insufficient.' (Quoted in Luxemburg, *op. cit.* p.286.)

[39] *Perepiska*, pp.127-29.

[40] Marx also often stressed this point, as can be seen from his statements cited on p.487 above.

and so the above contradiction will be constantly reproduced at a new, higher level . . . however, this dialectical study of the question is totally different to the abstract and hence extremely simplified ('linear') conception of the Russian Narodniks.

But how did Engels react to Danielson's letter? Did he deny the existence of the problem itself; did he declare it to be a simple 'misunderstanding', as did the Russian opponents of the Narodniks? Not at all. In fact he energetically stressed the point that according to Marx's theory the solution to the conflict between capital's unrestricted drive for valorisation and capitalist society's limited ability to consume was to be sought primarily (but not exclusively!) in the expansion of the capitalist economic order, in the creation of the internal market. But he conceded to his correspondent that this would be a contradictory and painful process – especially for a country such as Russia which started along the road of capitalist development at a relatively late stage, and had no significant external market at its disposal. For, as long as 'Russian industry is confined to the internal market, production can only cover internal consumption. And that can only grow slowly . . . Because it is one of the necessary accompanying phenomena of large-scale industry that it destroys its own market by precisely the process through which it creates it. It creates it by means of the destruction of peasant domestic industry. But the peasantry cannot live without domestic industry. They are ruined as peasants; their purchasing power is reduced to a minimum, and until they are established in new conditions of existence as proletarians, they furnish only a very poor market for the newly arisen factories.' However, the capitalist mode of production is 'full of internal contradictions' and the 'tendency to destroy its own internal market at the same time as it creates it is one of them'. Another contradiction is the 'hopeless position' to which capitalism leads in the last analysis 'and which comes about more quickly in a country without external markets (like Russia) than in countries which are more or less capable of competing on the open world market'. ('The latter', he added, 'can resort to the heroic means of trade policy, i.e. the forcible opening up of new markets.'[41])

Engels concluded his exchange with Danielson in a further letter with the words : 'I will gladly agree with you that insofar as Russia is the last country to be conquered by capitalist large-scale industry, and at the same time is also a country with an incomparably larger rural population than all other countries, the revolutionary change caused by the economic revolution must be much deeper and more

[41] *Perepiska*, pp.137-38.

Q

acute there than anywhere else. The process of the replacement of no less than 500,000 large landowners and approximately 80 million peasants can only be accomplished at the cost of terrible suffering and convulsions. History is the cruellest of goddesses, and she drives her chariot of triumph over mountains of corpses – not only in war, but also in "peaceful" economic development.'[42]

2. Bulgakov's and Tugan-Baranovsky's interpretation of Marx's analysis of extended reproduction

In contrast to Engels, the Russian adversaries of the Narodniks 'seized the bull by the horns'. Their main aim was to expose the theoretical shortcomings in the Narodniks' theory, and in so doing they skilfully used Marx's analysis of the reproduction of total social capital. However, as Rosa Luxemburg pointedly remarked, they proved 'too much' : 'The question was whether capitalism in general and Russian capitalism in particular was capable of development; these marxists however have proved this capacity to the extent of even offering us the theoretical proof that capitalism can go on for ever.'[43]

What is most immediately obvious in the essays and books of the Russian Legal Marxists in question is that they continually confuse Marx's abstract analysis with capitalist reality (as the Austro-marxists did later), and therefore draw conclusions from this analysis which are in no way justified.

This is evident even in the case of the most gifted and 'orthodox' of the Legal Marxists, Bulgakov.[44] Bulgakov of course realised that Marx's schemes 'look at neither the industrial cycle nor the periodically recurring crises, and that they consequently cannot accurately represent the real course of economic life. However, the point is that they in principle demonstrate the possibility of extended reproduction, and that this possibility will also become a reality.'[45]

Nonetheless, despite this reservation Bulgakov is convinced that these schemes, just as they are, offer the final and absolute solution

[42] *ibid.*

[43] Luxemburg, *op. cit.* p.325.

[44] We omit Struve here, whose boundless optimism as to the future outlook for Russian capitalism was criticised by Engels in a letter to Danielson. (See *MEW* Vol.39, pp.148-49.)

[45] S.Bulgakov, *O rynkach pri kapitalistitscheskom proizvodstve* (On the Question of Markets under the Capitalist Mode of Production), Moscow, 1897, p.165.

to the realisation problem. He writes, 'The most important questions of the theory of markets are resolved by the analysis of the exchange between the two departments' (Department I and Department II). 'Such an investigation shows where the market for surplus-value is to be found, how those parts of the commodity-product of the various capitals, which represent the value of the used-[up] constant capital, circulate, and how, finally, the consumption of the wages and surplus-value of those enterprises which produce non-consumable products is possible.'[46]

And in another passage of his book we read : 'The main difficulty in the analysis of the process of extended reproduction is to explain how the extension of production in Departments I and II is possible, although the first Department only produces constant capital and the second only variable. This difficulty is overcome by the fact that I accumulates constant capital for itself and for II, and II accumulates variable capital for itself and for I. Hence the difficulty of accumulation reduces itself to the exchange of those portions of the product which each department accumulates for the other.'[47]

So far, so good. In fact, Bulgakov overlooks the fact that the solution to the realisation problem, which the schemes offer, is only a solution at a quite abstract level and cannot, therefore, be a complete solution. Apart from this shortcoming there is little to object to in his formulation. Not satisfied with this, however, Bulgakov goes further. Because Departments I and II are exclusively dependent on one another in the schemes in Volume II, and require no other purchases, he outlines a grotesque picture of the absolute self-sufficiency of capitalist production – not only in the hypothetical world of the schemes, but also in reality ! He says, 'Mr. Tugan-Baranovsky was absolutely right in maintaining that capitalist production creates, simply by means of its growth, an uninterruptedly expanding market and that the degree of this expansion of the market depends solely on the presence of productive forces.'[48]

But why was Tugan right? Simply because, in Marx's schemes, Department I, which produces means of production, provides, from the first year, 'an independent demand' for the means of consumption from Department II, and Department II a similar demand for means of production from I : 'In this way a closed circle is already formed at the initial stage of capitalist production, in which it is totally independent of the external market, but is self-sufficient and

[46] *ibid.* pp.28-29.
[47] *ibid.*
[48] *ibid.* p.246.

in the position to grow automatically as it were, by means of accumulation.'[49]

And in one of the concluding chapters of his book Bulgakov even
asserts, in so many words, that 'the only market for the products of
capitalist production is this production itself', and that consequently,
'the only barrier to the extension of production lies in capital itself
and in its need for growth'.[50]

As we see, despite his marxist orthodoxy Bulgakov ends up
attributing a meaning to Marx's schemes which is not essentially
different from the harmonistic understanding expounded by Ricardo,
McCulloch and Say. But, could he reconcile this interpretation with
Marx's numerous remarks on the 'limited consuming power of the
masses' as the 'ultimate cause of all real crises'? Or, in other words,
how does the limited purchasing power of society affect the realisation
of products in general, and the realisation of surplus-value in
particular?

Bulgakov thought he had answered this question when he said:
'Consumption, the satisfaction of human needs, represents but an
incidental moment in the circulation of capital. The volume of
production is determined by the volume of capital and not by the
size of social requirements. Therefore the extension of production
does not have to be accompanied by a growth in consumption; an
antagonism can even exist between them . . . As we have seen,
capitalist production strives to reduce, relatively, both the share of
variable capital and the consumption fund of the capitalists . . .
Evidently the extension of capitalist production is primarily due to
Department I, that is, the production of constant capital, and only
a relatively small part can be credited to Department II, which
produces immediately for consumption.' And this alone, continues
Bulgakov 'shows well enough the role which consumption plays in
capitalist production and indicates where we should expect to find
the principal market for the commodities produced under capitalism'. Thus, 'capitalist production can expand without limit even
within the narrow limits of the profit motive and crises, independent
of consumption, and even when the latter is declining.' To be sure:
'Capitalist production pays with crises for its divergence from the
real aim of production. However, it is itself independent of consumption.'[51]

As can be seen, Bulgakov recognises that the significance of social

[49] *ibid.* p.210.
[50] *ibid.* pp.238, 259.
[51] *ibid.* pp.161-62.

consumption is expressed in periodically recurring economic crises. But he denies that these crises have anything to do with the realisation problem; according to him they are solely the result of the uneven development of the individual branches of production, and are therefore to be regarded as mere crises of disproportionality. For 'the fundamental and sole condition for the possibility of extended production consists in the proportionality of the different branches of industry. If this condition is adhered to, the level of production is then simply determined by the extent of capital accumulation, by the necessity of its growth.'[52] And everything which Marx had to say on crises is to be interpreted in this sense.

The question of foreign markets occupies a special place in Bulgakov's work. He confronts the 'fantastic dogma according to which capitalist production necessarily requires external markets', and emphatically denies that this is the case. As Luxemburg expressed it, his main argument is that the 'sceptics' from Sismondi to the Narodniks, 'obviously consider external commerce as a "bottomless pit" into which the internally produced, undisposable surplus of capitalist production disappears never to be seen again. Bulgakov, for his part, triumphantly points out that foreign trade is indeed not a pit, and certainly not a bottomless one, but rather appears as a double-edged sword, that exports always belong with imports, and both of them have to balance each other. Thus, whatever is pushed out over one border will be brought back over another, simply in a changed useful form. "We must find room for the commodities that have been imported as an equivalent for those exported within the bounds of the given market, and as this is impossible *ex hypothesi*, to have recourse to an external demand would only generate new difficulties." '[53]

At first sight Bulgakov's arguments appear very convincing; all the more so as he can still call on the well-known passage from Volume II of *Capital* where Marx declares the introduction of external trade into the analysis of the process of social reproduction to be irrelevant.[54] However, it should not be overlooked that in Volume II, as we have already pointed out several times, Marx studied the reproduction of social capital only in its 'basic form',[55] that is, at a quite abstract level. At this stage of the analysis the bringing in of external trade would only have served to 'confuse without

[52] *ibid.* p.158.
[53] Rosa Luxemburg, *op. cit.* p.309.
[54] See *Capital* II, p.474.
[55] *ibid.* pp.461, 463.

contributing any new element to the problem, or to its solution'.[56] However, as soon as one approaches the more developed concrete relations matters appear in a quite different light, as Rosa Luxemburg pointed out in her polemic against Bulgakov. Difficulties arise in the sphere of realisation which cannot be dealt with in the schemes in Volume II,[57] the very same difficulties which can be ameliorated, for a shorter or longer period of time by external trade.

Thus, imported commodities can be directly employed in the production process. So the 'change in the useful form' was nothing other than the act of realising the value of the exported commodities. (If, for example, English manufacturers export textiles and import cotton in return then at the same time this is *pro tanto* also the solution to their 'realisation problem', because by this they are able to transform a part of their accumulated surplus-value into the elements of production necessary for the expansion of their factories.) On the other hand Bulgakov himself admits, 'that partial overproduction of one or two commodities can be overcome if the surplus can be exported and sold on external markets. In this case external trade provides a safety-valve, which can protect any one given country from a partial or general crisis of overproduction'[58] (which means nothing more than that the country concerned finds a way of realising its surplus of commodities through external trade). Thus, Bulgakov's arguments lose all their persuasiveness when we turn to individual branches of production in particular countries in the concrete capitalist world.

After having denied any theoretical connection between the realisation problem and the question of external trade Bulgakov had to construct a specific theory of external trade which, as Luxemburg said, 'was not borrowed from Marx, but rather from the German scholars of bourgeois political economy'. In fact, 'there is no room in this theory for foreign trade. If capitalism forms a "closed circle" from the very beginning, if, chasing its tail like a puppy and in complete "self-sufficiency" it creates a limitless market for itself and can spur itself on to ever greater expansion, then every capitalist country must also be a closed and "self-sufficient" economic whole.'[59]

From this perspective 'the necessity of the external market for a capitalist country has causes which do not have their origin in the organisation of capitalist production itself, but which are external to

[56] *ibid.* p.474.
[57] See p.332 above.
[58] Bulgakov, *op. cit.* pp.200-01.
[59] Luxemburg, *op. cit.* p.306.

this organisation';[60] such causes can therefore only be of a 'historical' or 'geographical' nature![61] So, for example, England has to compensate for certain deficiencies which are due to its climate and the conditions of its soil. However, this does not apply to vast countries of a continental type, such as the USA or Russia, which can produce all – or nearly all – the necessary raw materials and foodstuffs themselves.[62] So it is hardly surprising that Bulgakov, in opposition to the Narodniks, predicted 'a great and brilliant future'[63] for Russian capitalism, still in its infancy, and even cherished the hope that Russia would soon be in a position to defeat its competitors on the world market[64] – a strange wish for a follower of Marx's theory. But Bulgakov was not the only one to indulge in such dreams.

Let us now turn to another opponent of the Narodniks – the Russian professor Tugan-Baranovsky who, although less gifted and original than Bulgakov, gained a big reputation in the West, and whose writings had a great influence on the thinking of German social-democratic theoreticians (Hilferding, O. Bauer). Rosa Luxemburg pointed out the difference in the approach of the two authors when she wrote : 'Like Bulgakov, Tugan-Baranovsky starts from Marx's analysis of social reproduction, which gave him the key to finding his whereabouts in this bewildering maze of problems. But while Bulgakov, the enthusiastic practitioner of Marx's theory, only sought to follow him faithfully and simply attributed his conclusions to the master, Tugan-Baranovsky, on the other hand, lays down the law to Marx who in his opinion did not know how to turn his brilliant exposition of the reproduction process to good account.' And in another passage : 'Bulgakov made an honest attempt to project Marx's scheme on to the real, concrete relations of the capitalist economy and capitalist exchange; he endeavoured to overcome the difficulties resulting from this . . . But Tugan-Baranovsky does not need any proof, he does not greatly exercise his brains. Because the arithemetical proportions come out satisfactorily and can be continued *ad infinitum*, this is to him proof that capitalist production can likewise proceed without let or hindrance – provided the said proportion continues to obtain.'[65]

We can add to this that Tugan-Baranovsky also loved to push his arguments to extremes and to indulge in paradoxes, which cer-

---

[60] Bulgakov, *op. cit.* p.260.
[61] *ibid.* p.183.
[62] *ibid.* pp.170-73.
[63] *ibid.* p.225.
[64] *ibid.* p.218.
[65] Luxemburg, pp.311, 315.

tainly did not add to the scientific value of his analyses. However, the conclusions of his first book do not differ fundamentally from Bulgakov's.[66] Like Bulgakov, Tugan also proclaimed the absolute self-sufficiency of capitalist production and its alleged independence from social consumption. Like Bulgakov he too denied that the spur to obtain external markets originated in the inherent laws of capitalism. And finally, he too derived economic crises, solely from the disproportionality between the various branches of industry. In all these aspects both authors can be regarded as forerunners of the later 'neo-Harmonist' trend in marxist economics. If we can discover a difference in their views this is to be found more in the accent given to different aspects – not what Tugan says, but how he says it.

A few passages from his book will suffice to illustrate this. 'The schemes quoted' (Tugan means here the schemes from Volume II as modified by him) 'must, as evidence, demonstrate the in-itself simple axiom that capitalist production creates a market for itself. If social production can be extended, if the productive forces are sufficient for this, then, with the proportional division of social production, demand too must undergo a corresponding expansion; for, under these conditions, every newly-produced commodity represents newly appearing purchasing power for the acquisition of other commodities.'[67] However, 'if the expansion of production has no practical limits, then we must assume that the expansion of the market is likewise unlimited, for if social production is organised proportionally there is no other barrier to the expansion of the market except the productive forces which society has at its disposal.'[68]

This already unwittingly constitutes an interpretation of Marx's schemes along the lines of Say's theory. But this is not the only thing which Tugan wants to derive from the schemes. According to him one can also derive the 'highly important conclusion that in capitalist society the demand for commodities is in a certain sense independent of the total volume of social consumption : this total volume of social consumption can decrease and at the same time aggregate social demand for commodities can grow – regardless of how absurd that might appear from a "commonsense" point of view. The accumula-

---

[66] M. von Tugan-Baranovsky, *Studien zur Theorie und Geschichte der Handelskrisen in England*, Jena 1901. (The author was unfortunately unable to obtain a copy of the first, 1894 edition, published in Russia. Rosa Luxemburg also used the later German translation, which certainly differs at numerous points from the first edition, as Tugan-Baranovsky had in the meantime become an open revisionist.)

[67] *ibid.* p.25.

[68] *ibid.* p.231.

tion of social capital leads to a contraction of the social demand for means of consumption, and simultaneously to an increase in the aggregate social demand for commodities.'[69]

This is, of course, the exact opposite of Marx's schemes, in which the advance of accumulation is accompanied by a steady growth in social consumption. In order to corroborate his hypothesis Tugan has to have recourse to one factor which was not considered in the schemes in Volume II, namely the law of the increasing organic composition of capital : 'Technical progress is expressed by the fact that the significance of the means of labour, machines, increases in comparison to living labour, to the worker himself . . . compared to the machine the worker recedes into the background, and at the same time the demand which arises from the worker's consumption is also placed in a position of less importance in comparison to the demand which arises from productive consumption by the means of production. The entire workings of the capitalist economy take on the character of a mechanism existing for-itself, as it were, in which human consumption appears as one simple moment of the process of the reproduction and circulation of capital.'[70]

And what does Tugan conclude from all this? Simply that 'machines have stepped into the place of living labour, the means of production have replaced the means of consumption as the market for commodities'. Thus, 'national income can fall and national demand simultaneously rise; the increase in national wealth can be accompanied by a reduction in national income, although that may sound paradoxical'.[71]

As we see, Tugan has succeeded in totally separating production from social consumption. It comes as no surprise that his fantasy later[72] led him to present a picture of capitalist society in which the entire working class has disappeared, with the exception of one single worker : and this single worker serves an enormous mass of machinery, which in turn produces new machines – without this situation leading to a discrepancy between production and social consumption.[73]

This is enough on the subject of Tugan-Baranovsky's 'marxism gone mad', which in fact only differs from Bulgakov's version in the extreme manner in which it is formulated, rather than in its funda-

[69] *ibid.* p.25.
[70] *ibid.* p.27.
[71] *ibid.* p.193.
[72] *Theoretische Grundlagen des Marxismus*, 1905.
[73] Quoted from W.Alexander, *Kampf um Marx*, 1932.

mentals.[74] However, Tugan and Bulgakov do differ on one point: their evaluation of Say's law. Whereas Bulgakov, in accordance with his orthodox marxism, was critical of Say, Tugan accepted the latter's thesis almost without any reservations whatsoever. He writes: 'For my part, I in no way assert that this theory is correct in every detail. Despite this I regard the kernel of the theory, its principal idea, namely that with a proportional division of social production the supply of commodities must coincide with demand – not only as correct, but as indisputable. In my opinion every objection which has been raised against this idea exhibits a serious lack of understanding of it!'[75]

However, it must be admitted that in this respect Tugan was merely more consistent than Bulgakov, since the 'harmonistic' interpretation of Marx's theory by the Legal Marxists was, at root, merely a revival of Classical 'optimism' in marxist guise. This indeed demonstrates the unexpected consequences which follow from the use of Marx's schemes of reproduction outside of their overall context, and taken in isolation.

III. LENIN'S THEORY OF REALISATION[76]

We characterised Bulgakov and Tugan-Baranovsky as forerunners of the later neo-Harmonist tendency in marxist economics. But doesn't this observation put us in a theoretical dilemma? It is well known that both authors had an ally in their controversy with the Narodniks in the young revolutionary marxist, Lenin, who

[74] 'Tugan-Baranovsky', wrote Hilferding, 'sees only the specific economic forms of capitalist production and overlooks the natural conditions which are common to all production, whatever their historical form, and thus comes to the strange idea of a form of production which only exists for production's sake whilst consumption appears as a burdensome appendage. If this is "madness", it does have "method", and in fact marxist method, for this analysis of the historical specificity of capitalist production is specifically marxist. It is marxism gone mad, but still marxism, which makes Tugan's theory appear at the same time to be so strange and interesting.' (*Finanzkapital*, p.355, Note 1.)

[75] Tugan-Baranovsky, *op. cit.* p.27. It can be seen from Bulgakov's polemic against Tugan's first book that he proposes the same idea there too.

[76] If we pay particular attention to Lenin's writing on the problem of realisation, this is not just because of their unquestionable theoretical significance, but also because extensive extracts from them are usually appended to editions of *Capital*, apparently to serve as a kind of official exegesis of Marx's work. This practice began in the 1930s, and Lenin would certainly not have tolerated it if he had been alive.

shared many of their views on this subject. Does this mean that Lenin also has to be attributed with a harmonistic interpretation of Marx's economic theory? Bulgakov and Tugan-Baranovsky both left the socialist movement after a few years and became ideologues for the liberal Russian bourgeoisie. However, a scientific theory cannot be judged by the political careers of its advocates, and the later ideological development of Bulgakov and Tugan-Baranovsky is therefore no more relevant than that of Lenin in this respect.

So the dilemma we have just referred to is perhaps not as great as it might appear at first sight. We should remember that economic and sociological theories do not exist in the ether of pure knowledge, but also, as a rule, fulfil certain social functions. Looked at in this way, Lenin's theoretical alliance with the Legal Marxists does not appear quite so remarkable.

At that time the Russian marxists saw it as one of their main tasks to put up a determined opposition to the ideology of the Narodniks, which denied the special historical role of the Russian working class and wanted to drag the socialist movement of the country back onto the utopian path of a specifically Russian agrarian socialism. In order to overcome this ideology the theoretical assumptions on which it was based had to be shown to be without foundation. Thus, when the Narodniks spoke of the basic impossibility of the realisation of surplus-value in the capitalist economy, and proved this by reference to the external markets which the Russian bourgeoisie lacked, to declining social consumption and to the crises of overproduction inherent in capitalism, then their marxist opponents sought to prove that the realisation of surplus-value was also possible without foreign markets and even with a low level of consumption, and that consequently the phenomenon of crises of overproduction should not be derived from the difficulty of realisation but from the anarchy of the capitalist economic system. The abstract analysis of the hypothetical conditions for equilibrium in the process of expanded reproduction in a 'pure' capitalism was supposed to provide adequate proof of this. It is hardly surprising then that the marxist opponents of the Narodniks far overestimated the scope of this analysis, and occasionally interpreted it in a way which could scarcely be reconciled with the real meaning of Marx's theory.

From this perspective the pointed remark which Plekhanov inserted into the second edition of his pamphlet *Our Differences*, in which he distanced himself both from the Legal Marxists and from Lenin, is understandable. He wrote, 'I never subscribed to those theories of the market in general, and of crises in particular which overcame our Legal Marxist literature like an epidemic in the nineties. According to this theory, whose chief propagandist can be regarded as Tugan-Baranovsky, reproduction has no barriers, and crises are simply explained by the disproportion of the means of production. Werner Sombart regarded Tugan-Baranovsky as the father of this supposedly new theory. However, the real father of this by no means new theory was J.B.Say in whose *Treatise* it is elaborated in detail. Apart from Tugan-Baranovsky this theory is also propagated by V. Lenin in both his *Remarks on the Question of the Theory of Markets* (1899), and his book *The Development of Capitalism in Russia*.'[77]

Plekhanov's critique of Lenin, the polemical sharpness of which is mainly attributable to the factional struggles inside Social Democracy at that time, is certainly exaggerated. It does seem to contain an element of truth, however, especially if Lenin's earliest writings on the question of realisation are considered. For example, we find the following sentence in his first work *On the So-called Question of Markets* : 'The market is simply the expression of the division of labour in the commodity economy, and therefore its growth is as limitless as that of the division of labour.'[78]

This is an assertion which lends itself to an interpretation similar to that of the optimistic perspective expounded by the Classical economists. There are also several passages dealing with crises in Lenin's significant work *A Characterisation of Economic Romanticism* which are equally questionable, where he agrees with Ricardo, and even McCulloch, in their dispute with Sismondi.[79]

However, if we turn to Lenin's later writings, and in particular to those cited by Plekhanov, we must reject the view that there is a fundamental similarity between Lenin's and Say's conception. These works do nonetheless contain some one-sided arguments and exaggerated formulations which we ought to deal with here.

Lenin was of course right when he referred the Narodniks to the fact that their doubts as to the possibility of the realisation of surplus-value could be answered by Marx's analysis of the social process of reproduction. In fact Marx showed in his schemes how, if definite proportions are maintained in the exchange between the production goods and consumption goods industries, capitalist society can not only renew its constant and variable capital, but also enlarge

[77] Cited from Kowalik, *The Economic Theory of Rosa Luxembourg* in the journal *Ekonomista*, 1963, No.1. [In Polish.]

[78] Lenin, *Collected Works* Vol.I, p.100.

[79] Lenin later clearly changed his mind on this, as can be seen from his marginal notes on Luxemburg's *Accumulation of Capital* (in *Leninskii Sbornik* XXII, p.357).

it by means of the capitalisation of a portion of surplus-value. Lenin referred his Narodnik adversaries to these schemes several times, and adds : 'Once these basic propositions', the basis of Marx's schemes of reproduction, 'are taken into account, the problem of the realisation of the social product in capitalist society no longer presents any difficulty'. And further on : 'By establishing these main theoretical propositions Marx fully explained the process of the realisation of the product in general, and of surplus-value in particular in capitalist society.'[80]

But the question then arises : can the proof of the possibility of the realisation of surplus-value in principle, as offered by Marx's analysis, actually count as the 'complete explanation' of the realisation problem? In fact, the analysis in Volume II consciously disregards such decisive factors in the real capitalist world as the growth in the organic composition of capital, and the increase in relative surplus-value, i.e. factors which continually disturb the balance between production and consumption and consequently must place even greater obstacles in the way of the realisation of the social product.

Along with the Legal Marxists, Lenin thought that this objection could be countered by reference to the relatively faster growth in the industries producing the means of production. Thus, he stresses : 'On the problem of interest to us here, that of the home market, the main conclusion from Marx's theory of realisation is the following : capitalist production, and consequently, the home market, grow not so much on account of articles of consumption as on account of means of production. In other words, the increase in means of production outstrips the increase in articles of consumption.'[81]

And in another passage : 'This disparity' of production and consumption, 'is expressed, as Marx demonstrated clearly in his schemes, by the fact that the production of the means of production can and must outstrip the production of articles of consumption.'[82]

In actual fact, however, Marx's schemes show nothing of the kind, since, in both examples in Volume II, Department II develops at exactly the same speed as Department I. (This does not of course reflect concrete reality, but is a feature of the numerical examples chosen by Marx.) Hence, Lenin's thesis cannot be proved by reference to the schemes in Volume II, and like Tugan and Bulgakov before him, he is forced to connect the analysis of the reproduction

---

[80] Lenin, *Collected Works* Vol.3, pp.52, 68.

[81] *ibid.* p.54.

[82] Lenin, 'Reply to Mr. Nezhdanov', *Collected Works* Vol.4, p.162.

process in Volume II of *Capital*, with the law of the increasing composition of capital which was expounded in Volume III. He continually refers to the fact that, 'according to the general law of
capitalist production constant capital grows faster than variable . . .
Consequently the department of social production which makes the
means of production must grow faster than the one which produces
the means of consumption. Thus, for capitalism, the growth of the
internal market is to a certain extent "independent" of the growth
of individual consumption.' Of course, continues Lenin, 'the
development of production (and consequently of the home market)
chiefly on account of means of production seems paradoxical and
undoubtedly constitutes a contradiction. It is real "production as an
end-in-itself", the expansion of production without a corresponding
expansion of consumption. But it is a contradiction not of doctrine,
but of actual life.' For this is the feature 'that corresponds to the
historical mission of capitalism and to its specific social structure. The
former consists in the development of the productive forces of society;
the latter rules out the utilisation of these technical achievements by
the mass of the population.'[83]

The passages quoted here seem to be particularly characteristic
of Lenin's interpretation of the theory of realisation. It is clear that
his views on this were formed in the context of the specific situation
of Russia during the early period of capitalism, in which the industrialisation of the still semi-feudal country seemed to in fact offer
an unlimited market for the means of production. From this point
of view Lenin's conception is certainly correct for all those countries
which are at the stage of their industrial revolution and have yet to
create the bases for modern industry – a transport system and
mechanised agriculture – which normally takes place at the cost of
a very low level of subsistence for the mass of the population. And
when Lenin stressed the necessity and progressive nature of this
process, he demonstrated his profound sense of historical reality and
proved himself far superior to his Narodnik adversaries. But was he
justified in extrapolating this hypothesis, which was based on a particular historical situation, to capitalism in all its phases? It is quite
clear that capitalism must complete vast numbers of factories and
machines, railways, harbours etc. while it is building up its industrial
base, and that this process provides a rapidly growing market for the
means of production, stretching over decades. However, sooner or
later the basic phase of industrialisation is completed, and the industrial apparatus which has been created must produce goods for

[83] Lenin, *Collected Works* Vol.3, p.56.

individual use. Then the problem of the purchasing power of the masses enters the foreground and this cannot be evaded – unless one believes in 'Mr. Tugan-Baranovsky's carousel',[84] i.e. in his fantastic conception of 'the production of machines for their own sake'.

So much on Lenin's attempt to project the law of the increasing composition of capital onto Marx's schemes of reproduction and to interpret these schemes as implying a necessarily more rapid growth of the production-goods industries. We saw previously that the schemes of reproduction, which assume a parallel and equal development of Departments I and II, do not permit such an interpretation; and we now see that Lenin's thesis can only be valid for a limited period of time – namely the epoch of initial industrialisation – and can therefore in no way be regarded as a universal developmental law of capitalism.

However, this is not the only objection which can be raised against Lenin's interpretation of the schemes of reproduction. What appears even more questionable to us is the fact that he regarded the abstract analysis in Volume II as the final and definite word of the marxist theory of realisation, and accordingly did not wish to recognise the relevance of the later enlargements and modifications to this analysis, as found in Volume III. This creates considerable theoretical difficulties for him, which are mainly attributable to a certain misconceived 'orthodoxy' as far as Marx's writings are concerned.

This is evident in his polemic against Tugan-Baranovsky. As a revisionist and follower of Say, Tugan could not accept many passages in Volume III of *Capital* which contradicted his harmonistic interpretation of Marx's schemes. Above all, he attacked the well-known section in which Marx writes : 'The conditions of direct exploitation, and those of realising it are not identical . . . The first are only limited by the productive power of society, the latter by the proportional relation of the various branches of production *and*[85] by society's power of consumption.'[86]

Tugan-Baranovsky interprets this passage to the effect that, according to Marx 'proportionality . . . on its own does not guarantee the possibility of marketing the products. The products may not find a market even if the distribution is proportional – that is apparently the meaning of Marx's above-quoted words.' This is an inter-

---

[84] This is how Luxemburg characterised Tugan's notion of the accumulation of capital.

[85] Author's italics.

[86] *Capital* III, p.244.

pretation which one must agree to as Marx's formulation does not in fact permit any other.

However, Lenin denies this. He replies: 'No; that is not the meaning of these words. There are no grounds for seeing in them some sort of a correction to the theory of realisation expounded in Volume III. Marx is here merely substantiating that contradiction of capitalism which he indicated in other places in *Capital*, that is the contradiction between the tendency towards the unlimited expansion of production, and the inevitability of restricted consumption (as a consequence of the fact that the mass of the population are proletarian). Mr. Tugan-Baranovsky will of course not dispute the fact that this contradiction is inherent in capitalism; and since Marx points to it in the passage quoted we have no right to look for some other meaning in his words.'[87]

The question is then: what other meaning? One different to the one which Tugan-Baranovsky attributes to the schemes of reproduction in Volume II – namely that the realisation of the social product depends exclusively on the proportionality between the various branches of production? In this instance Lenin's polemic misses its target: since, instead of attacking Tugan's harmonistic interpretation of Marx's theory of realisation as a mere 'theory of proportionality', he seems to sanction this interpretation – simply attempting to give it a 'further' modified form. In his view, ' "the consuming power of society" and "the proportional relation of the various branches of production" are not conditions that are isolated, independent of and unconnected with each other. On the contrary, a certain level of consumption is one of the elements of proportionality.'[88]

As far as Lenin's interpretation of the concept of proportionality is concerned, it is indisputable that any disturbance of the balance between production and consumption sooner or later also brings about a disturbance in the proportionality of the various branches of production. On the other hand it is clear that the concept of proportionality – if thought out to its conclusion – must also comprise the mutual correspondence of production and consumption. However, it in no way follows from this that the concepts of 'proportionality', and 'equilibrium of consumption and production' cannot be separated from one another, or that they should always be regarded as equivalent. Thus, for example, Marx deduced partial crises precisely from the disproportionality of the various branches of production,

[87] Lenin, *Collected Works* Vol.4, p.58.
[88] *ibid.* p.58.

without regard to the relation of production and consumption.[89] Lenin too wrote in an article directed against Struve : 'Marx's theory not only does not restore the apologetic bourgeois theory, but on the contrary, provides a most powerful weapon against apologetics. It follows from this theory that even with an ideally smooth and proportional reproduction and circulation of the aggregate social capital, the contradiction between the growth of production and the narrow limits of consumption is inevitable.'[90]

But if this is the case, then Tugan's interpretation of Volume III was correct; it must then be admitted that according to Marx the realisation of the social product does not only depend on the 'proportional relation between the various branches of production', but also on 'society's capacity to consume', and it is difficult to see what could be the theoretical use of Lenin's concept of 'proportionality in a further sense'.

And Lenin's oft-repeated argument that Marx merely 'stated' the contradiction between production and consumption in the cited passages from *Capital*, and 'nothing more', is even less convincing.[91] This contradiction plays a key role in Marx's theory and was only left out of consideration in the analysis in Volume II for methodological reasons which we have yet to deal with.

It is evident that Lenin's postulate, according to which the relation of production and consumption is to be subsumed under the concept of proportionality, brings him uncomfortably close to Bulgakov's and Tugan's 'disproportionality theory' of crises. We read in his book *The Development of Capitalism in Russia* : 'If one speaks of the "difficulties" of realisation of crises etc. which arise from them, one must admit that these "difficulties" are not only possible, but necessary . . . Difficulties of this kind, due to the disproportion in the distribution of the various branches of industry, constantly arise, not only in realising surplus-value, but also in realising variable and constant capital; in realising not only the product con-

---

[89] Cf. *Theories* II, p.521 : 'It goes without saying that, in the whole of this observation, it is not denied that too much may be produced in individual spheres and therefore too little in others; partial crises can thus arise from disproportionate production . . . and a general form of this disproportionate production may be overproduction of fixed capital, or on the other hand, overproduction of circulating capital . . . However, we are not speaking of crisis here insofar as it arises from disproportionate production, that is to say, the disproportion in the distribution of social labour between the individual spheres of production . . . Ricardo etc. admit this form of crisis.'

[90] Lenin, *Collected Works* Vol.4, p.87.

[91] Lenin, *Collected Works* Vol.2, pp.168-69, Vol.4, pp.58-59.

sisting of articles of consumption, but also that consisting of means of production.'[92]

And even more clearly in another passage : 'The irregular production of an unrealisable product (i.e. crisis) is inevitable in capitalist society, as a result of the disturbance in the proportion between the various branches of industry.' ('But', Lenin adds, 'a certain level of consumption is one of the elements of proportionality'.)[93] This essentially amounts to the disproportionality theory of crises, even if this theory is modified in such a way that the proportionality of the branches of production now also depends on the relations of consumption.

Lenin's interpretation of Marx's theory of realisation also explains his complete rejection of Rosa Luxemburg's book *The Accumulation of Capital* (1912), which we will deal with later. Thus, in March 1913, he wrote to the publishers of the Russian journal *Sotsialdemokrat*, which was published in Paris : 'I have just read Rosa's new book . . . She has got into a shocking muddle. She has distorted Marx. I'm very glad that Pannekoek, Eckstein and Otto Bauer have all condemned her book with one accord and used the same arguments which I already used against the Narodniks in 1899. I intend to write about Rosa in Number 4 of *Prosveshchenie*.'[94]

Unfortunately, Lenin never wrote the planned article. However, in his well-known essay on Marx, which was published in the Russian encyclopaedia *Granat* in 1915, we find the following bibliographical note : 'Marx's theory of the accumulation of capital is dealt with in a new book by R. Luxemburg. An analysis of her incorrect interpretation of Marx's theory can be found in Eckstein's discussions in *Vorwärts*. See also O. Bauer's article in *Neue Zeit* and Pannekoek's in the *Bremer Bürgerzeitung*.'[95]

Apart from the abnormally severe tone of Lenin's letter, which can largely be explained by the factional disputes at this time between the Bolsheviks and the 'Social Democracy of the Kingdom of Poland and Lithuania' (SDKPL), led by Luxemburg, what is noticeable about Lenin's comments is his complete agreement with the criticism of Rosa Luxemburg by the Austro-marxists Bauer and Eckstein. (The article by the Dutch leftist Pannekoek does not carry any theoretical weight). But what is the source of this curious theoretical agreement between the spokesman of the most radical wing of marxism and

[92] Lenin, *Collected Works* Vol.3, p.47.
[93] Lenin, *Collected Works* Vol.4, p.161.
[94] Lenin, *Collected Works* Vol.35, p.94.
[95] Lenin, *Collected Works* Vol.21, p.90.

such exponents of social democratic 'neo-Harmonism' as Bauer and Eckstein? This situation certainly requires an explanation.

In this connection we should remember that in the years before the outbreak of the First World War Lenin's political sympathies lay more with Kautsky's 'Centre' than with the German 'Left' led by Luxemburg.[96] However, what is of interest to us is not the political, but the theoretical background to Lenin's attitude to Luxemburg's book. This background was clearly stated by Lenin himself, in the letter quoted above, in 1913. He adopted a position of solidarity with the Austro-marxist critics of Rosa Luxemburg for precisely the reason that their views coincided with the arguments he had put forward against the Narodniks in 1899; and he rejected Luxemburg's book not only because of its erroneous criticism of Marx's reproduction schemes, but also because its theoretical conception ran so counter to the version of the realisation theory which he himself had proposed. And it is evident that he still adhered to the version which he had defended alongside the Legal Marxists in the 1890s.[97]

However, the methodological aspect of the question is perhaps still more important. When the young Lenin wrote his treatise on the realisation problem neither Marx's *Theories* nor the *Grundrisse* were known to him : he could have had only a less than adequate insight into the methodologically very complex structure of Marx's economic work. We now know that according to Marx's plan for the

---

[96] Lenin wrote to Shlyapnikov on 27 November 1914: 'Rosa Luxemburg was right when she wrote, long ago, that Kautsky has the "subservience of a theoretician" – servility in plainer language, servility to the majority of the Party, to opportunism.' (Lenin, *Collected Works* Vol.35, p.169.)

[97] We should mention in this context that Lenin never attacked the fundamental basis of Bulgakov's or Tugan's ideas. On the contrary, he defended these ideas against their critics, and even recommended his readers 'who do not find it possible to become conversant with Volume II of *Capital* to study the presentation of Marx's theory of realisation in Bulgakov's book'. Lenin did admittedly criticise Tugan from time to time, but only because of his 'deviations' from Marx, and because he claimed that a contradiction existed between Volumes II and III of *Capital*. But even after this polemic he defended Tugan, Bulgakov and himself, against Struve's criticism that they had all derived the 'harmony of production and consumption' from Marx's schemes. He wrote: 'In my opinion, Struve's polemic against the above-mentioned writers is due not so much to an essential difference of views as to his mistaken conception of the content of the theory he defends . . . Neither Marx nor those writers who have expounded his theory and with whom Struve has entered into a polemic deduced the harmony of production and consumption from this analysis, but, on the contrary, stressed forcefully the contradictions that are inherent in capitalism and that are bound to make their appearance in the course of capitalist realisation.' (Lenin, *Collected Works* Vol.4, p.74.)

structure of the work, the first two volumes only figure as the analysis of *'capital in general'* and that consequently the results which Marx obtained in these volumes – although extraordinarily important – still had to be concretised and supplemented by a later stage of the analysis, that of *'capital in concrete reality'*. The early marxists, including Lenin, understandably overlooked this. It is no great surprise that Lenin rather exaggerated the theoretical validity and relevance of the analysis in Part III of Volume II of *Capital*, and tended to regard it as Marx's 'last word' on the theory of realisation. This also explains his attempts to reconcile the results of this analysis in a literal and scholastic fashion with the numerous passages in Volume III, which apparently contradicted it, and which Tugan and the Narodniks were so happy to rely on for their case.[98] (Tugan, so that by comparing the 'true' Marx of Volume II with the 'erring' Marx of Volume III, he could interpret the schemes of reproduction from Volume II in such an uncompromisingly harmonistic sense; the Narodniks, so that they could attribute Sismondi's underconsumption theory of crises to Marx – despite the schemes.) Nevertheless, in actual fact the exposition in Volume III in no way contradicts that in Volume II (Lenin was certainly correct on this point). It does, however, represent a further stage of the analysis; a stage at which the question is no longer that of the conditions for the equilibrium of the capitalist economy in its 'normal' course, but that of the causes of the necessary disturbances to this equilibrum, i.e. the analysis of crises and the tendency to breakdown which is inherent to capitalism. What follows from this is that the schemes of reproduction and the analysis of Volume II can in no way, on their own, offer the 'complete explanation' of the realisation problem, but can only do

[98] Additional proof is furnished by a quote from Lenin's polemic against Danielson. Danielson had used a note written by Marx and inserted by Engels – 'Note for future amplication' in Chapter XVI of Volume II of *Capital* – as proof of his case. Lenin writes: 'After the words quoted above, the note goes on to say: "However, this pertains to the next part", i.e. to the third part. What is this third part? It is precisely the part which contains a criticism of Adam Smith's theory of the two parts of the aggregate social product . . . and an analysis of the "reproduction and circulation of the aggregate social capital" i.e. of the realisation of the product. Thus, in confirmation of his views, which are a repetition of Sismondi's, our author quotes a note that pertains "to the part" which refutes Sismondi: "to the part" in which it is shown that the capitalists *can* realise surplus-value, and that to introduce foreign trade in an analysis of realisation is absurd.' (Lenin, *Collected Works* Vol.2, p.169.)

In fact, the entire remark would have been untenable if Lenin had known that Marx's reference had not referred to the 'next part' as meaning Section III of Volume II, but the 'Section on Competition' as envisaged in the original outline, which corresponded to the later Volume III of *Capital*.

this in connection with Marx's theory of crisis and breakdown. And it seems to us that the greatest deficiency in Lenin's theory of realisation is that he overlooked this fundamental fact.[99]

IV. HILFERDING'S INTERPRETATION OF MARX'S SCHEMES OF REPRODUCTION

We saw that Marx's analysis of the social reproduction process primarily assisted the Russian marxists of the 1890s in demonstrating the possibility and inevitability of capitalist development in Russia, in opposition to the Narodnik 'sceptics'. The situation was different in Germany and Austria, however, where this analysis was interpreted by the official theorists of social democracy to mean that capitalism could be extended without limit, and that it was not threatened by any kind of breakdown conditioned by its inner laws.

In fact, Hilferding, the most famous economist of the Austro-marxist school, wanted to read practically everything possible into the schemes in Volume II ! Not only that if the reproduction of total social capital is to take place normally definite proportions have to be maintained between Departments I and II, production-goods' and consumption-goods' industries, which is self-evident from the schemes – but rather also that 'capitalist production and reproduction both on a simple and extended scale, can proceed undisturbed, *only* if these proportions are kept to'.[100] (As if proportionality were the only condition on which undisturbed reproduction depended!) But there is still more to come. Hilferding continues : 'It therefore in no way follows that the crisis has to have its origin in the under-consumption of the masses, which is inherent in capitalism. Nor does it follow from the schemes in themselves, that there is the possibility of a general overproduction of commodities : rather it is shown that any extension of production that can in fact take place with the available forces of production is possible.'[101]

Hilferding is of course right about the last point; the possibility of overproduction does not follow from the schemes 'in themselves', because they only, in fact, investigate the conditions for a normal, undisturbed course of reproduction. However, since the impossibility

[99] It was not until after this chapter was completed that the author saw the interesting essay *Rebels and Renegades* (1946) by the American socialist Paul Mattick, in which the criticism made here of Lenin's theory of realisation is to some extent anticipated.

[101] *ibid.*

[100] Rudolf Hilferding, *Das Finanzkapital*, 1927, p.318.

of overproduction also does not follow from these schemes, it is difficult to see what purpose is served by referring to them at all, i.e. what
conclusions can be drawn from them as far as the actual capitalist
world is concerned.

It naturally does not occur to Hilferding to deny the empirical
facts of overproduction and the underconsumption of the masses, or
the role which these facts play as moments in actual crises. What he
is aiming at with his observation of the schemes 'in themselves' is
something different : that in the last analysis only the proportional
relation of the individual branches of production is decisive in the
process of social reproduction. His disproportionality theory of
crises, as well as his rejection of any breakdown theory, then follows
quite consistently from this.

But let us look more closely at how Hilferding furnishes the
evidence for this : he states at the beginning of the section on crises
in his book, 'The expression "overproduction of commodities" in
general says as little as the expression "underconsumption". One can
only speak of underconsumption, understood strictly, in a physiological sense; by contrast, the expression has no meaning in economics, where it can only mean that society has consumed less than it
has produced.[102] However, it is not easy to see how that can be
possible if production has taken place in the correct proportions. Since
the aggregate product is equal to constant capital plus variable
capital plus surplus-value, $(c + v + s)$, both v and s must be consumed, and the elements of constant capital used up must be replaced.
It follows that production can be extended indefinitely, and this will
never lead to the overproduction of commodities, i.e. to a situation
in which more commodities, which in this context and for this standpoint means more use-values, more goods, are produced than can
be consumed.'[103]

This is really a strange argument! Almost everything which
Marx wrote on crises was intended to prove that it was just this
periodically recurring overproduction which represented the 'basic
phenomenon of crises',[104] and that this overproduction has its 'ulti-

[102] The illogicality of the concepts 'overproduction' and 'underconsumption' did not prevent Engels from characterising capitalism as an economic
system which 'produces a far greater quantity of means of subsistence and
means of development than capitalist society can consume because it keeps the
great mass of the real producers artificially away from these means of subsistence and development.' (Letter to Lavrov, 12 November 1875, *Selected
Correspondence*, p.285.)
[103] Hilferding, *op. cit.* p.300.
[104] *Theories* II, p.528.

mate cause in the poverty and restricted consumption of the masses'.[105] But now we discover that although this might apply in the crude world of fact, it does not apply in the 'for-itself' world of the schemes, and that consequently the expressions 'overproduction' and 'underproduction' have no meaning in political economy . . . but why?

Simply because the imaginary society in general, which Hilferding substitutes 'in this context' for the real world, could never produce too many use-values, goods, and because it is also in its power to eliminate any shortage of articles of consumption by a more proportional division of production! Of course, the concept of 'underconsumption' could only have a 'physiological meaning' for such a society. However, we are not discussing physiology here, but an economy – and not merely an economy in general, but a capitalist economy. That is, not society 'as such', but a class society, 'in which the mass of producers remains more or less restricted to necessities . . . in which consequently this great majority of producers remains more or less excluded from the consumption of wealth – insofar as wealth goes beyond the bounds of the necessary means of subsistence'[106] (which means nothing other than that they have 'underconsumed'); and in which, on the other hand, the ruling class (as distinct from the ruling classes of previous epochs) likewise have to subject their consumption to the drive for valorisation, i.e. to fulfil the role of a 'producer of overproduction'.[107] For this reason periodic crises of overproduction must occur in this society – even with the most perfect proportionality of the branches of production, and it is impossible to see what we could gain theoretically, by renaming these crises 'crises of disproportionality', or by equating the overproduction of commodities with the overproduction of 'goods'.

Let us leave to one side Hilferding's theory of crisis, which in our opinion merely represents another version of the critique offered by the Ricardian school[108] on the theory of crises of overproduction. Our concern is Hilferding's thesis that – 'as the schemes show' –

105 *Capital* III, p.484.
106 *Theories* II, p.528.
107 *Theories* I, p.283.
108 In Marx's critique of Ricardo's theory of crises we read: 'The word overproduction itself leads to error. So long as the most urgent needs of a large part of society are not satisfied, or only the most immediate needs are satisfied, there can of course be absolutely no talk of an overproduction of products – in the sense that the amount of products is excessive in relation to the need for them. On the contrary, it must be said that on the basis of capitalist production, there is constant underproduction in this sense . . . But overproduction of products and overproduction of commodities are two entirely different things. If Ricardo thinks that the commodity-form makes no

'capitalist production could be extended indefinitely'. How does he prove this thesis? By presenting us with Marx's scheme for simple reproduction, where not only 'do the elements of constant capital used replace each other', but also 'where both v, and s are consumed' – i.e. where it is impossible to speak of a problem of the realisation of surplus-value!

Perhaps Hilferding made a slip of the pen; perhaps he actually meant the scheme for extended reproduction, and simply forgot to say that s is consumed individually as well as industrially? It is certainly consumed in this way in Marx's schemes. However, the fact that the cited scheme can be continued indefinitely does not mean that actual capitalist production 'could be extended indefinitely'. Hilferding simply neglects the fact that the schemes of reproduction in Volume II deliberately disregard technical progress i.e. the increase in the organic composition of capital, increase in the rate of surplus-value etc., and that the introduction of any one of these aspects would completely overturn them! This is convincing evidence of the kind of absurdities to which a confusion of the abstract schemes with actual capitalist reality must lead, especially, if like Hilferding, one tries to prove, on the basis of these schemes, that the idea of an economic breakdown of capitalism 'is in no way a rational idea'.[109] Marx's schemes of reproduction are simply a tool of analysis, and can-

---

difference . . . then this is in fact in line with his presupposition that the bourgeois mode of production is the absolute mode of production, hence it is a mode of production without any definite specific characteristics.' (*Theories* II, p.527.) And, further: 'All the objections which Ricardo and others raise against overproduction etc. rest on the fact that they regard bourgeois production either as a mode of production in which no distinction exists between purchase and sale . . . or as social production, implying that society, as if according to a plan, distributes its means of production and productive forces in the degree and measure which is required for the fulfilment of the various social needs . . . This explanation of overproduction in one field by underproduction in another field therefore means merely that if production were proportionate, there would be no overproduction. The same could be said if demand and supply corresponded to each other, or if all spheres provided equal opportunities for capitalist production and its expansion . . . i.e. if all countries which traded with one another possessed the same capacity for production (and indeed for different and complementary production). Thus overproduction takes place because all these pious wishes are not fulfilled.' (*ibid.* pp.528-29, 532.)

It is clear how much Hilferding's theory coincides, in its fundamentals, with Ricardo's.

[109] We refer here to the last chapter of Hilferding's work, in which he discusses the approaching breakdown of the imperialist policy of finance capital, 'which will be a matter of a political and social breakdown, not an economic breakdown, which is in no way a rational idea'. (*op. cit.* p.47.)

not therefore be used for such a purpose. As an alternative to Hilferding's disproportionality theory of crises we would like to cite a number of Marx's own comments on the subject of crises, which deal with the antithesis of production and consumption.

Thus Marx writes against Ricardo in the chapter on crisis in Volume II of *Theories* : 'He overlooks the fact that the commodity has to be converted into money. The demand of the workers does not suffice, since profit arises precisely from the fact that the demand of the workers is smaller than the value of their product, and that it [profit] is all the greater the smaller, relatively, is their demand.' In the long run 'the demand of the capitalists amongst themselves is equally insufficient . . . Overproduction arises precisely from the fact that the mass of the people can never consume more than the average quantity of necessities, that their consumption therefore does not grow correspondingly with the productivity of labour.'[110] This is because 'the mere relationship of wage-labourer and capitalist implies :

1. that the majority of the producers (the workers) are non-consumers (non-buyers) of a very large part of their product, namely of the means of production and the raw material;

2. that the majority of the producers, the workers, can consume an equivalent for their product only so long as they produce more than this equivalent, that is, so long as they produce surplus-value, or surplus-product. They must always be *overproducers*, produce over and above their needs, in order to be able to be consumers or buyers within the limits of their needs.'[111]

In another passage in the *Theories* we read : 'The whole process of accumulation in the first place resolves itself into *production on an expanding scale*, which on the one hand corresponds to the natural growth of the population, and on the other hand forms an inherent basis for the phenomena which appear during *crises*. The criterion of the expansion of production is *capital* itself, the existing level of the conditions of production and the unlimited desire of the capitalists to enrich themselves and enlarge their capital, but by no means *consumption*, which from the outset is inhibited since the majority of the population, the working people, can only expand their consumption within very narrow limits, whereas the demand for labour, although it grows *absolutely*, decreases relatively, to the same extent as capitalism develops.'[112]

Finally, from the same part of the *Theories: Overproduction* is

110 *Theories* II, p.468.
111 *ibid.* p.520.

specifically conditioned by the general law of the production of capital : to produce to the limit set by the productive forces, that is to say, to exploit the maximum amount of labour with the given amount of capital, without any consideration for the actual limits of the market or the needs backed by the ability to pay; and this is carried out through continuous expansion of reproduction and accumulation, and therefore constant reconversion of revenue into capital, while on the other hand, the mass of the producers remain tied to the average level of needs, and must remain tied to it according to the nature of capitalist production.'[113]

Marx remarks in the same sense in one of the manuscripts of *Capital* : 'Contradiction in the capitalist mode of production : the workers as buyers of commodities are important for the market. But as sellers of their own commodity – labour-power – capitalist society tends to keep them down to the minimum price. Further contradiction; the periods in which capitalist production exerts all its forces regularly turn out to be periods of overproduction, because production potentials can never be utilised to such an extent that more value may not only be produced but also realised; but the sale of commodities, the realisation of commodity capital and thus of surplus-value, is limited, not by the consumer requirements of society in general, but by the consumer requirements of a society in which the vast majority are always poor and must always remain poor.'[114]

However, the contradiction which we are discussing is pointed out most acutely in a passage in Volume III which has already been cited : 'The conditions of direct exploitation, and those of realising it, are not identical. They diverge not only in place and time, but also logically. The first are only limited by the productive power of society, the latter by the proportional relation of the various branches of production and the society's power of consumption. But this last-named is not determined either by the absolute productive power, or by the absolute consumer power, but by the consumer power based on antagonistic conditions of distribution, which reduce the consumption of the bulk of society to a minimum varying within more or less narrow limits. It is furthermore restricted by the tendency to accumulate, the drive to expand capital and produce surplus-value on an extended scale.' Consequently : 'the more productive power develops, the more it finds itself at variance with the narrow basis on which the conditions of consumption rest.'[115]

[113] *ibid.* pp.534-35.
[114] *Capital* II, p.320.
[115] *Capital* III, pp.244-45.

In another passage we read : 'Since the aim of capital is not to minister to certain wants, but to produce profits, and since it accomplishes this purpose by methods which adapt the mass of production to the scale of production, not vice versa, a rift must continually ensue between the limited dimensions of consumption under capitalism and a production which forever tends to exceed this immanent barrier.'[116] This is because 'as matters stand, the replacement of the capital invested in production depends largely upon the consuming power of the non-producing classes while the consuming power of the workers is limited partly by the laws of wages, partly by the fact that they are used only as long as they can be profitably employed by the capitalist class. The ultimate reason for all real crises remains the poverty and restricted consumption of the masses as opposed to the drive of capitalist production to develop the productive forces as though only the society's absolute capacity for consumption constituted their limit !'[117]

The above passages (to which many more could be added)[118] at least show the large role which Marx attributed to the contradiction between production and consumption, as the reason for crises of overproduction – although he was himself an opponent of the traditional 'underconsumption theory'. The following passage from *Capital* shows how he also, on the other hand, rejected the so-called disproportionality theory of crises : 'To say that there is no general

[116] *ibid.* p.256.
[117] *ibid.* p.484.
[118] Cf. for example *Theories* III, p.120: 'Ricardo here equates "productively" and "profitably", whereas it is precisely the fact that in capitalist production "profitably" alone is "productively", that constitutes the difference between it and absolute production, as well as its limitations. In order to produce "productively", production must be carried on in such a way that the mass of the producers are excluded from the demand for a part of the product. Production has to be carried on in opposition to a class whose consumption stands in no relation to its production – since it is precisely in the excess of its production over its consumption that the profit of capital consists.' And elsewhere : 'The fact that bourgeois production is compelled by its own immanent laws, on the one hand, to develop the productive forces as if production did not take place on a narrowly restricted social foundation, while, on the other hand, it can develop these forces only within these narrow limits, is the deepest and most hidden cause of crises, of the crying contradictions within which bourgeois production is carried on and which, even at a cursory glance, reveal it as only a transitional, historical form. This is grasped rather crudely but none the less correctly by Sismondi, for example, as a contradiction between production for the sake of production, and distribution, which makes an absolute development of productivity impossible.' (*Theories* III, p.84.)

overproduction, but rather a disproportion within the various branches of production, is no more than to say that under capitalist production the proportionality of the individual branches of production springs as a continual process from disproportionality . . . It amounts furthermore to demanding that countries in which capitalist production is not developed should consume and produce at a rate which suits the countries with capitalist production. If it is said that overproduction is only relative, this is quite correct; but the entire capitalist mode of production is only a relative one, whose barriers are not absolute. They are absolute only for this mode, i.e. on its basis. How could there otherwise be a shortage of demand for the very commodities which the mass of the people lack, and how would it be possible for this demand to be sought abroad, in foreign markets, to pay the labourers at home the average amount of necessities of life? This is only possible because in this specific capitalist interrelation the surplus-product assumes a form in which its owner cannot offer it for consumption, unless it first reconverts itself into capital for him . . . In short, all these objections to the obvious phenomena of overproduction (phenomena which pay no heed to these objections) amount to the contention that the barriers of *capitalist* production are not barriers of *production generally*, and therefore not barriers of this specific, capitalist mode of production. The contradiction of the capitalist mode of production, however, lies precisely in its tendency towards an absolute development of the productive forces, which continually come into conflict with the specific *conditions* of production in which capital moves, and alone can move.'[119]

So much then, on the so-called disproportionality theory.

V. ROSA LUXEMBURG'S CRITIQUE OF MARX'S THEORY OF ACCUMULATION

1. The historical and methodological background

Our discussion of Hilferding has illustrated how Germany's official marxist theory made use of the schemes of reproduction in Volume II. Although this theory seemed both radical and 'orthodox' in fact it simply led to a rejection of the theory of breakdown and to a Vulgar-Economic explanation of crises as mere crises of disproportionality. That is, completely in the spirit of Tugan and the

[119] *Capital* III, p.257.

Russian Legal Marxists! Rosa Luxemburg's book *The Accumulation of Capital*, whose central theme – disregarding the secondary and subsidiary material – involves stressing the idea of breakdown and hence the revolutionary kernel of marxism, can only be understood against this background. That is, as a reaction to the neo-harmonist interpretation of Marx's theory.

But why did this task fall to Rosa Luxemburg rather than Lenin? The explanation lies in the different historical situation of Russian and German marxism. In contrast to Russian marxists of the 1890s, whose theoretical interests were constrained by their struggle against Narodnik ideology and who were compelled to prove the viability of a Russian capitalism which was still in its infancy, Rosa Luxemburg lived and was politically active in a country in which capitalism had not only reached the summit of its power, but was already showing clear signs of its future decline; furthermore, her adversaries were not the disciples of a utopian peasant socialism, but a powerful workers' bureaucracy, strongly rooted in the masses, which despite its 'marxist' principles located itself squarely within the prevailing social order and hoped to achieve all its demands for social and political progress within its confines. Thus, whereas in Russia at the turn of the century it was still necessary to stress the inevitability and historical progressiveness of capitalist development, the task of the marxist left in Germany was just the opposite – to give prime place to the idea of the inevitable economic and political breakdown of the capitalist social order. Rosa Luxemburg's book was intended to fulfil precisely the latter theoretical task.

However, it in no way follows from this that we accept Rosa Luxemburg's specific theory of accumulation, according to which capital accumulation can only be explained by having recourse to the so-called 'third person', i.e. exchange with the non-capitalist milieu;[120] or that we regard her critique of Marx's schemes of reproduction as correct. On the contrary : it is regrettable that Luxemburg was only able to defend the concept of breakdown in the extreme form of a basically incorrect critique of Marx's theory of reproduction. It would nevertheless be pedantic to re-examine this critique,

---

[120] Of course, Marx was obliged to disregard the role of the 'third person', and factors extraneous to capital in general in his abstract analysis of the process of accumulation. This is why Luxemburg's critique is erroneous here. However, this does not mean that the 'third person' has to be left out at further stages of the analysis, as many of Luxemburg's opponents assumed, quite incorrectly. On the contrary, the actual process of the accumulation of capital can scarcely be understood with taking account of these factors.

which has long been recognised as incorrect, and whose main defect consisted in the fact that Luxemburg, without noticing it, continually fell back on the presuppositions of simple reproduction in the analysis of extended reproduction. What is more important, and educative, is to examine the reasons which led her to do this. Henryk Grossmann seems to have hit on the right idea when he wrote : 'It is to Rosa Luxemburg's great historical credit that – in conscious opposition and protest against attempts at distortion by the neo-Harmonists – she adhered to the basic idea of *Capital* and attempted to support it by demonstrating an absolute economic limit to the further development of the capitalist mode of production.' But, instead 'of examining Marx's reproduction scheme in the context of Marx's entire system, and his theory of accumulation, in particular . . . she unintentionally succumbed to the influence of those she wanted to oppose, i.e. she believed that Marx's scheme did in fact permit a limitless accumulation "ad infinitum in a circle – as in Tugan-Baranovsky's theory".' And because she considered that 'the possibility of limitless accumulation ad infinitum actually resulted from Marx's schemes of reproduction, and that Tugan, Hilferding and later Otto Bauer had *correctly* deduced this notion from the schemes, she abandoned it in order to save the concept of breakdown which originated in Volume I of *Capital*.'[121]

In our opinion Grossmann here goes far towards explaining Rosa Luxemburg's mistakes. In addition to this, however, her incorrect interpretation of the schemes of reproduction seems to have its roots in an insufficient understanding of the methodology of Marx's work. It is of course true, as Lukacs remarked, that Luxemburg was a 'genuine dialectician',[122] and this explains the great theoretical satisfaction which a study of her work provides. Despite this she clearly underestimated the so-called 'Hegelian inheritance' in Marx's thought,[123] and was therefore not entirely conscious of the real structure of his work. We have already dealt with her confusion of the distinction between individual capital and total social capital,

[121] Henryk Grossmann, *op. cit.* pp.20, 280-82.

[122] Lukacs, *History and Class Consciousness*, p.182. (Cf. the interesting essay by L.Basso, 'Rosa Luxemburg: The Dialectical Method', in *International Socialist Journal*, November 1966.)

[123] It may well have been the product of a passing mood and a feeling of annoyance at the sham orthodoxy of her critics when she wrote to her friend Diefenbach from prison on 8 March 1917 : 'This [i.e. simplicity of expression] is generally to my taste, which, as in art or science, values only what is simple, peaceful and generous, which is why, for example, the famed Volume I of *Capital* with its Hegelian Rococo ornamentation is quite abhorrent to me at

with the much more important distinction between 'capital in general' and 'capital in reality' elsewhere,[124] and we do not therefore have to return to it here. We also know that she incorrectly compounded total social capital with capital as it concretely, historically exists. In her view, Marx's concept of a 'pure capitalist society' could only be of use in the study of the production and circulation process of individual capital; this concept would lose all meaning as soon as one turned to capitalist society as a whole, and in particular to the problem of the accumulation of the total social capital.

In other words, Rosa Luxemburg also misunderstood the role allotted to the model of a pure capitalist society in Marx's work. She did not grasp that it represented a heuristic device, intended to help in the illustration of the developmental tendencies of the capitalist mode of production, free from 'all disturbing accompanying circumstances'.[125] (From this standpoint the endless discussions as to the actual historical possibility of a pure capitalist society were completely irrelevant.) The methodological intent of this procedure is clear. If, even under the strictest assumptions, i.e. in the abstract model of a pure capitalist society, it is possible for surplus-value to be realised and for capital to accumulate – within certain limits – then there is no theoretical need to have recourse to external factors, such as foreign trade, the existence of a third person, or state intervention. In this sense Marx's model completely stood the test. And because Rosa Luxemburg overlooked this fact, she also neglected to realise that the results of the analysis of reproduction in Volume II could only be of a provisional nature, i.e. they needed to be supplemented at a later, more concrete, stage of the analysis.

Rosa Luxemburg's methodological error must seem all the more surprising in that she came very near to a correct understanding of the methodological assumptions behind Marx's schemes when she

the moment (which merits, from the party standpoint, 5 years' imprisonment and 10 years' loss of rights).' (Luxemburg, *Briefe an Freunde*, p.85.) However, this remark does show that Rosa Luxemburg sometimes overlooked the dialectical content hiding behind Marx's 'Hegelian style'.

[124] See pp.183ff above.

[125] 'In considering the essential relations of capitalist production', wrote Marx in the *Theories*, 'it can therefore be assumed that the entire world of commodities, all spheres of material production . . . are (formally or really) subordinated to the capitalist mode of production, for this is happening more and more completely, since it is the principal goal . . . On this premise – which expresses the limit of the process and which is therefore constantly coming closer to an exact presentation of reality – all labourers engaged in the production of commodities are wage-labourers, and the means of production in all these spheres confront them as capital.' (*Theories* I, pp.409-10.)

wrote : 'The premises which are postulated in Marx's diagram of accumulation accordingly represent no more than the historical tendency of the movement of accumulation and its final theoretical result. The process of accumulation strives everywhere to substitute simple commodity economy for natural economy, and the capitalist economy for the simple commodity economy, and to establish the exclusive and universal domination of capitalist production in all countries and for all branches of production.'[126]

And in Rosa Luxemburg's *Anti-Critique* we read : 'Marx himself never dreamed of presenting his own mathematical models as any sort of *proof* that accumulation was in fact possible in a society consisting solely of capitalists and workers. Marx investigated the internal mechanism of capitalist accumulation and established certain economic laws on which the process is based. He started roughly like this : if the accumulation of gross capital, that is, in the entire class of capitalists, is to take place, then certain quite exact quantitative relations must exist between the two large departments of social production : the production of the means of production and the production of means of consumption. Progressive expansion of production and, at the same time, progressive accumulation of capital – which is the object of it all – can only proceed unhindered if such relations are maintained so that the one large department of production continuously works hand-in-hand with the other. Marx sketched a mathematical example, a model with imaginary numbers, to illustrate his thoughts clearly and exactly, and he uses it to show that if accumulation is to proceed, then the individual points in the model (constant capital, variable capital, surplus-value) must behave in such and such a way to each other.'[127]

But if this is correct, if Marx's model was simply a tool for showing the conditions for equilibrium in an expanding capitalist economy in their pure form, then Rosa Luxemburg's assertion that it represents a 'bloodless abstraction' cannot be upheld; it simply proves that her critique of Marx's schemes of reproduction was also methodologically without foundation.

2. *The schemes of reproduction and technical progress*

We do not wish, however, to discuss merely what was defective in Rosa Luxemburg's critique : it also has its positive sides, which have mostly gone unmentioned by her detractors.

[126] Luxemburg, *op. cit.* p.417.
[127] Luxemburg, *Anti-Critique*, pp.68-69.

We mean by this her pointing out of the fact, already known to us, that Marx's schemes of extended reproduction disregard all those changes in the mode of production which are caused by technical progress – namely, the increasing organic composition of capital, the increase in the rate of surplus-value, and the rising rate of accumulation. As soon as one attempts to introduce these changes the conditions for equilibrium are disturbed, and the formula cII $+ \beta$ cII $=$ v I $+ \propto$ I $+ \beta$ v I can no longer be employed.

Consider the following numerical example based on Tugan-Baranovsky's scheme for reproduction, which is intended to illustrate the extended reproduction of capital.

$$\text{I} \quad 840\,c + 420\,v + 210\,\alpha + 140\,\beta\,c + 70\,\beta\,v$$
$$\text{II} \quad 600\,c + 300\,v + 150\,\alpha + 100\,\beta\,c + 50\,\beta\,v$$

This diagram corresponds to the general formula for equilibrium as,

$$600\,c\,\text{II} + 100\,\beta\,c\,\text{II} = 420\,v\,\text{I} + 210\,\alpha\,\text{I} + 70\,\beta\,v\,\text{I}$$

However, as soon as we increase the organic composition of the capital accumulating in each department from 2 : 1 to 3 : 1 we obtain the following result :

$$\text{I} \quad 840\,c + 420\,v + 210\,\alpha + 157.5\,\beta\,c + 52.5\,\beta\,v$$
$$\text{II} \quad 600\,c + 300\,v + 150\,\alpha + 112.5\,\beta\,c + 37.5\,\beta\,v$$

In this instance a commodity surplus is produced in Department II, which can no longer exchange its 600 $+ 112.5\,\beta$ c $= 712.5$ units of value without a remainder for 420 v $+ 210\,\alpha + 52.5\,\beta$ v $= 682.5$ units, but is left with an undisposable remainder of commodities amounting to 30 units. This corresponds to the fact that with a rising organic composition of capital fewer workers are taken on, and therefore social consumption cannot be expanded sufficiently to absorb the entire commodity-product of Department II.

Similar disturbances necessarily arise if the rate of surplus-value rises or if a larger portion of the newly created surplus-value is accumulated than in previous periods of production. In such cases the smooth progress of extended reproduction, as envisaged in the scheme, becomes impossible as the disproportion in the relations of exchange between the two departments, which comes about as a consequence of technical progress, must explode their previous proportionality.

We see then that 'if we take into consideration technical changes

R

in the mode of production, in the context of the advance of accumulation . . . it cannot come about without completely disrupting the basic relations of Marx's scheme'.[128] Rosa Luxemburg is undoubtedly right on this point. However, it cannot be concluded from this 'failure' of the schemes of reproduction (as she supposed), that accumulation is completely 'impossible', but simply that any revolution in the productive forces which takes place on a social scale must bring the given state of equilibrium of the branches of production to an end and lead, via all kinds of crises and disturbances, to a new temporary equilibrium. Therefore the results of Rosa Luxemburg's critique are only the necessary limits of the area within which Marx's schemes are valid – schemes which are deliberately confined to the investigation of the relations of equilibrium of extended reproduction under constant conditions of production and which must therefore disregard all the moments which alter these conditions. But if one still wanted to introduce the changes in the mode of production which result from the rising productivity of labour, this would only prove how the hypothetical conditions of the normal course of reproduction 'change into so many conditions of abnormal movement, into so many possibilities of crisis',[129] which in no way belonged to the tasks of the analysis in Volume II of *Capital*.

However, doesn't this underestimate the importance of this analysis? Not at all. It is clear that Marx's model of extended reproduction in a condition of equilibrium in pure capitalism was not supposed to be a true reflection of the concrete capitalist world, nor could it be. For one thing, it leaves to one side the anarchy of production which rules in actual capitalism, and furthermore it takes no account of the conflict between production and consumption which is inseparable from the essence of capitalist production. Consequently, the proportional development of the various branches of production, and the equilibrium between production and consumption, can only be obtained, in this mode of production, in the midst of continuous difficulties and disturbances. Naturally, this equilibrium must at least be attained for short periods of time, or else the capitalist system could not function at all. In this sense, however, Marx's schemes of reproduction are in no way a mere abstraction, but a piece of economic reality, although the proportionality of the branches of production postulated by these schemes can only be temporary, and 'spring as a continual process from disproportionality'.[130]

---

[128] Luxemburg, *Accumulation of Capital*, p.339.
[129] *Capital* II, p.499.
[130] See the quotation from *Capital* on p.490 above.

3. The neo-Harmonist applications of the schemes

The fact that the formula for the equilibrium of extended reproduction, which forms the basis of Marx's schemes, only applies to accumulation under constant conditions of production, induced several authors to undertake painstaking mathematical work in order to show that the 'failure' of this formula does not follow from the formula itself, but results from the far too strict assumptions to which Marx tied his schemes; and that consequently, with an appropriate modification of these assumptions, a scheme of extended reproduction could be constructed, which, even taking into consideration technical progress, would exhibit a permanent equilibrium between the two departments of social production. What lurks behind all these attempts – although its original proponents were not always aware of this – is the desire to present the reproduction and accumulation of capital as an automatic and permanent process, which does not encounter any of the barriers which originate in the nature of the capitalist mode of production, and which could not, therefore, lead to an economic breakdown of this mode of production.

The best-known example of this kind is the scheme of reproduction set out by Otto Bauer in his critique of Rosa Luxemburg,[131] the intention of which is to prove the possibility of an undisturbed progress of accumulation, even with a constantly rising organic composition of capital. In order to accomplish this, Otto Bauer (like Tugan-Baranovsky before him) must drop one of the basic assumptions of Marx's schemes : namely the assumption that the only relation between Departments I and II is the mutual exchange of their respective products. Instead he allows Department II, which is always left with an undisposable remainder of commodities as a result of the technical changes caused by the rising organic composition of capital, to 'invest' each year a sum of money corresponding to the value of this remainder in Department I, so that the latter extends its production and buys the real remainder of commodities in the next year. In this way both departments of social production are able to grow and accumulate without a discrepancy arising in the value of the products to be exchanged by them and without the perpetual motion of capital accumulation threatening to come to a standstill.

These are the main points of Bauer's method. It is clear that his numerical example only apparently represents a further development of Marx's scheme of reproduction. For he could have demons-

[131] Otto Bauer, *'Die Akkumulation des Kapitals'*, in *Die Neue Zeit*, 1913.

trated what he wanted to prove just as well by means of an industrial combine which sets up a subsidiary agricultural-industrial concern in order to provide the combine's workers and capitalists with the necessary means of subsistence. In the book-keeping of the complex the auxiliary plant could figure as 'Department II', which regularly 'invests' a part of its surplus-value in the main plant and 'exchanges' the means of subsistence produced by it for the machines from the main plant. However, such 'investments' and 'exchanges' would be of a purely fictitious nature and it is difficult to see what such calculations could contribute to the understanding of the actual process of reproduction in the real capitalist world.[132]

But isn't Otto Bauer's scheme of reproduction more 'realistic' than Marx's? Isn't it in fact true that in actual capitalist society portions of the surplus produced in one particular department are constantly transferred to other branches of production to be invested there? Shouldn't we therefore regard Otto Bauer's procedure as a considerable improvement on Marx's?

This is clearly the opinion of the Polish political economists Otto Lange and T.Kowalik. The latter writes : 'As far as the transfer of accumulation from one department to the other is concerned, history has without doubt proved Otto Bauer to be correct, for in economic practice capital is transferred both in its material and in its money-form.' Thus 'a considerable portion of social production can be used alternatively in the role of means of production as well as for the aim of personal consumption', and this fact is confirmed by the 'practice of the socialist countries, where' (Kowalik quotes Lange here) 'accumulation primarily takes place in Department II, but the main part of this accumulation is invested in Department I.'[133]

We can disregard here Kowalik's naïvely-empiricist attitude ! He seems to believe that questions of pure theory – such as the question of the hypothetical equilibrium in the capitalist society of Marx's diagrams ! – can be decided by reference to the practice of 'the socialist countries' (or any practice). What can be said on his argument itself is this : as far as the transfer of capital in its material form

---

[132] The reader will recall the objection raised by Luxemburg against Marx's schemes of reproduction, according to which 'accumulation in Department II is completely dominated and determined by that in Department I'. Although this is wrong in relation to Marx, it is correct in relation to Bauer, as Department II in his scheme does appear to be no more than a mere appendage to Department I and merely serves the constant expansion of the latter.

[133] T.Kowalik, *op. cit.* p.208.

is concerned Kowalik has clearly overlooked the fact that products which can be employed either as means of production or as means of consumption are excluded from Marx's schemes from the outset. We read on this subject in *Capital* II : 'Nor does it change matters if a part of the products of II is capable of entering into I as means of production. It is covered by a part of the means of production supplied by I, and this portion must be deducted on both sides at the outset, if we wish to examine in pure and unobscured form the exchange between the two large classes of social production, the producers of means of production and the producers of articles of consumption.'[134]

Thus the 'transfer of capital in material form' cannot get us over the difficulties raised by Rosa Luxemburg. And of equally little help is the transfer of capital in money-form, which underlines Bauer's method – regardless of how much it corresponds to the everyday practice of the capitalists – since this transfer has been practised since time immemorial, without any regard to the changes caused by technical progress. Therefore, in methodological terms, there is no reason at all to introduce it if the formula for the equilibrium of extended reproduction seems to fail; that is, when the issue is that of the difficulties of realisation which arise as a result of the increase in the organic composition of capital! Why did Marx himself not resort to the mode of capital transfer recommended by Otto Bauer, instead of drawing up schemes with such complicated quantitative relations in the two Departments? The answer is simple : because he wanted to use these schemes to show how the antinomy of use-value and exchange-value can be, and is, resolved at a social level. However, this can only be shown if the industries making production-goods and consumption-goods are considered as completely autonomous departments of social production which can only obtain their respective products by means of exchange, and only in this way accomplish the social change of form and matter.

This is enough on the methodological deficiences of the solution to the problem proposed by Otto Bauer. Our primary interest here is the question as to whether he is able to prove what he wants to prove with the aid of his method – namely the possibility of an unlimited accumulation of capital.[135] A moment's reflection

---

[134] *Capital* II, p.525.

[135] Of course Bauer denies that his schemes have this goal. He writes : 'This presentation cannot be regarded as an apology for capitalism, for whereas the apologists would want to prove the limitlessness of accumulation – that consumer power grows automatically with production – we on the other hand reveal the limit which is set to accumulation.' (*Die Neue Zeit,* 1913, p.887.)

shows that this attempt is bound to fail. That is, if the scheme for extended reproduction is constructed under the assumption of a constantly rising organic composition of capital then sooner or later one must arrive at a completely unrealistic and economically meaningless hypertrophy of Department I, i.e. the production of the means of production. Otto Bauer does precisely this : In order to express the growth in the organic composition of capital he allows society's variable capital to grow by 5% per year, but constant capital to grow by 10% annually. These differing rates of growth come about because the organic composition of newly-accumulated capital in his scheme is considerably higher than capital which has been invested previously. (And because, besides this, Bauer assumes the rate of surplus-value to be constant, he must add an ever larger portion of surplus-value to the accumulation fund.)

Bauer begins with the following diagram, which, for the sake of clarity, we express in Bukharin's symbols.

	c	v	s		
			α	βc	βv
I	120,000 +	50,000 +	37,500 +	10,000 +	2,500 = 220,000
II	80,000 +	50,000 +	37,500 +	10,000 +	2,500 = 180,000
	200,000 +	100,000 +	75,000 +	20,000 +	5,000 = 400,000

The general formula for equilibrium is in keeping with this diagram since 80,000 c + 10,000 β c = 90,000 units of value of Department II can be exchanged for 50,000 v + 37,500 α + 2,500 β v = 90,000 units of value from Department I.

Despite this the capitalists would be in a predicament if they wanted to invest the surplus-value obtained in the first year in the proportions shown above in the same departments in which it was

---

However, if one searches for this 'limit to accumulation' it turns out that he only means 'the tendency for accumulation to adjust itself to the increase in population'. He writes: 'The increase of productive capital in the country itself always remains limited by the growth of the available working population: variable capital can never grow faster than the population, and constant capital can only grow more rapidly than variable capital in a proportion determined by the level of development of the productive forces.' (*ibid.* pp.871-72.) However, if this is the case, if the accumulation of capital collides only with the barrier of 'available working population' which shows itself temporarily in the prosperity phase of the industrial cycle, then this accumulation can proceed into eternity, and Bauer's disavowal of apologetics is simply a pious wish.

produced, as we would then obtain the following product values in the second year :

	c	v	α	s βc	βv	
I	130,000 +	52,500 +	39,375 +	10,500 +	2,625	= 235,000
II	90,000 +	52,500 +	39,375 +	10,500 +	2,625	= 195,000
	220,000 +	105,000 +	78,750 +	21,000 +	5,250	= 430,000

However, in this case 90,000 c II + 10,500 β c would equal 100,500 units of value, whereas Department I would only have 52,500 v + 39,375 α + 2,625 β v = 94,500 units of value to transfer. Thus, an undisposable remainder of commodities would remain in Department II – a remainder which would get larger every year, and which would eventually lead to a crisis in the disposal of the product.

This does not occur in Bauer's scheme, however, because the capitalists in Department II invest a portion of their surplus-value in Department I, instead of proceeding with the production process according to the above diagram. Bauer says they can do this either by setting up new factories to produce means of production, or by buying up the shares in existing factories. In fact according to Bauer's calculation the production process in the second year must be continued in the following value composition, after the capital transfers from Department II to Department I :

	c	v	α	s βc	βv	
I	134,666 +	53,667 +	39,740 +	11,244 +	2,683	= 242,000
II	85,334 +	51,333 +	38,010 +	10,756 +	2,567	= 188,000
	220,000 +	105,000 +	77,750 +	22,000 +	5,250	= 430,000

If the above quantitative relations are established by transfers of capital the general formula for equilibrium can once more be applied, since 85,334 c II + 10,756 βc II = 96,090, and 53,667 v I + 39,740 α I + 2,683 β v I = 96,090. And since Bauer from this point onwards allows the capitalists of Department II annually to invest their excess surplus-value in Department I, at first sight it seems as if the numerical example could carry on *ad infinitum*.

In reality, however, this is nothing other than Tugan's 'carousel' – namely the production of machines for its own sake. Not unexpectedly the production of the means of production increases enorm-

502 · *The Making of Marx's 'Capital'*

ously rapidly in Bauer's diagram – whereas the production of consumption goods only increases slowly. Admittedly Otto Bauer's
scheme only continues for four years; but Henryk Grossmann took
the trouble to extend the same scheme up to 35 years. And in the 20th
year the following value composition for the aggregate product of
Departments I and II is the result:

$$1,222,252 \text{ c} + 252,691 \text{ v} + 117,832 \, \alpha + 122,225 \, \beta c + 12,634 \, \beta v = 1,727,634.$$

These figures show that the total social product has grown to
1,727,634, of which, however, only 383,157 are intended for human
consumption, whereas the other 1,344,477 have to be thrown back
into production as constant capital! And all this simply to secure the
undisturbed disposal of the capitalists' commodities and a frictionless course for Bauer's scheme! . . .

Such a hypertrophy of the production of means of production,
without a corresponding increase in social consumption, as necessarily
follows from Bauer's scheme, is surely incompatible with the spirit
of Marx's theory. Marx pointed out that 'constant capital is never
produced for its own sake but solely because more of it is needed in
spheres of production whose products go into individual consumption'.[136]

It is sufficient to confront this passage with Bauer's scheme to
see to what extent it in fact amounts to the same thing as Tugan's
'carousel'.[137]

It could naturally be objected that the extremely rapid pace
at which the production-goods industries overtake the consumption-
goods industries in Bauer's scheme is to be attributed to the unrealistic
rates of growth of 10% in Department I and 5% in Department II
which Bauer assumes. However, if one insists on constructing a
scheme which is supposed to reflect the rising organic composition of
capital, and despite this to show a frictionless course for capitalist
accumulation, the first department must be allowed to grow faster
than the second; so that – even if a smaller discrepancy between the
rates of growth of the two departments was to be assumed than that
in Bauer's scheme – one would attain the same absurd result,
although after a longer period of time.

[136] *Capital* III, p.305.
[137] As we discover from Kowalik's dissertation, Bauer's essay containing
the scheme has been printed several times in the Soviet editions of Luxemburg's works – evidently to serve as a kind of antidote. For example, in the
1934 Edition, pp.339-58.

But this is not all. Already in those few years for which Bauer continues his scheme it becomes clear that the growing organic composition of capital leads to the progressive fall in the rate of profit. In Bauer's example the profit rate has already fallen from 33.3% to 30.3% in the fourth year (s : (c + v)). It is now an easy task for Grossmann to prove mathematically that on the basis of Bauer's assumptions the capitalist system would have to break down in Year 35, because the relation of surplus-value to capital employed would have fallen so much that the capitalist class could no longer accumulate!

At this point we should remember that in Bauer's example the growth of relative surplus-value, which is supposed to accompany the increased organic composition, is not considered. However, can't the fall in the rate of profit be compensated for by the increase in relative surplus-value? As we already know from our study of the *Grundrisse* this question has to be answered in the negative. Marx refers there to the fact that the increase of relative surplus-value cannot be extended indefinitely, since, with technical progress, not only does the paid portion of the working day fall (and it has to fall) but so too does the relation of total living labour to the labour objectified in the means of production.[138] Therefore Bauer's scheme must eventually exhibit a progressive fall in the rate of profit – and with this lead to a collapse of the capitalist system – even if he had taken the rising rate of surplus-value into account.

What this shows in fact is that if one tries to replace Marx's model of extended reproduction by a model which takes into consideration as many factors as possible from capitalist reality, one very soon comes up against the barriers which are set to capitalist production by the nature of capital itself. It is not all surprising that Henryk Grossmann could use Bauer's scheme to prove that a tendency to breakdown is inherent in capitalism.[139] (In this sense Otto Bauer's scheme of reproduction can be characterised as his contribution – if unintended! – to the 'breakdown theory'.)

Conclusion

What is the result of our investigation? It is enough here to confine ourselves to a brief resumé of the foregoing.

The first conclusion which emerges from the decades-long

[138] Cf. p.409 above.
[139] See Grossmann, *Das Akkumulations- und Zusammenbruchsgesetz des kapitalistischen Systems*, 1929.

dispute over Marx's schemes of reproduction is clearly that these schemes should in no way be seen as a mere 'torso', as a theoretical experiment which Marx was unable to 'complete' because of a lack of time. On the contrary, everything indicates that Marx himself never intended to go beyond the form of the schemes of reproduction as they are in Volume II of *Capital* and that it is therefore senseless to expect more than they can actually accomplish. We have stressed several times that Marx's schemes only deal with the hypothetical conditions for the equilibrium of extended reproduction with constant conditions of production, and yet, despite their abstractness, represent a 'piece of economic reality'. Of course, in the actual capitalist world the extended reproduction and accumulation of capital is accomplished 'through a progressive qualitative change in its composition, i.e. through a constant increase of its constant component, at the expense of its variable component',[140] and this process is accompanied by the uneven extension of the domain of relative surplus labour i.e. by the increase in the fate of surplus-value. One should not forget however that this constant change in the mode of production is 'constantly interrupted by periods of rest, during which there is a merely quantitative extension . . . on the existing technical basis', by means of 'intermediate pauses in which accumulation works as simple *extension* of production'.[141]

And it is for such 'pauses' that the schemes of reproduction in Volume II are valid, showing the possibility of extended reproduction through the mutual adjustment of the production-goods and consumption-goods industries, and hence also the possibility of the realisation of surplus-value. However, all this could have been demonstrated without it being necessary also to include in the analysis of Volume II the factor of technical progress, which is expressed in the increase in the composition of capital.

However, couldn't Marx have gone further and elaborated the conditions for equilibrium of extended reproduction under the assumption of a constantly changing mode of production? We think we have shown that this was not possible, and the unsuccessful attempts at a solution by Tugan-Baranovsky and Otto Bauer only confirm this view. For, as soon as one attempts to introduce technical

[140] *Capital* I, p.781 (629).

[141] *ibid.* pp.578, 782 (450), (629). A similar view is expressed in the *Theories*: 'During the examination of reproduction, it is, in the first place, assumed that the method of production remains the same and it remains the same, moreover, for a period while production expands. The volume of commodities produced is increased in this case, because more capital is employed and not because capital is employed more productively.' (*Theories* I, p.522.)

progress into the schemes of reproduction, the conditions for equilibrium of reproduction turn into conditions for the disturbance of equilibrium, and any scheme which tries to get around this obstruction must turn out to be an economically insignificant 'mathematical exercise'. And this result, for which we have to thank Rosa Luxemburg, is irrefutable.

The second important result to which our investigation has led consists in the finding that the schemes of reproduction of Volume II merely represent one phase – although an extremely important one – in Marx's analysis of the social reproduction process, and that they consequently need to be supplemented by Marx's theory of breakdown and crises. What follows from this is that these schemes can only be understood in the total context of Marx's theory. (Here too, the concept of the totality proves itself to be methodologically crucial.) In fact the disturbances to the equilibrium of reproduction brought about by technical progress seem, in the first instance, only to prove that the course of capitalist production must repeatedly lead to crises, and consequently to the replacement of the prevailing temporary equilibrium by a new, equally temporary, equilibrium.

However, in reality they prove even more; namely that the contradictions of the capitalist mode of production, which are expressed in just these disturbances and in the tendency of the rate of the profit to fall, which they accelerate, are reproduced at a constantly higher level until finally the spiral of capitalist development reaches its end. And in this respect the apparently scholastic dispute over the interpretation of Marx's reproduction schemes must be regarded as positive and theoretically fruitful, despite all its errors and false conclusions.

31.
The Problem of Skilled Labour

I. BÖHM-BAWERK'S CRITIQUE

1. The problem of 'skilled' or 'complex' labour has probably been one of the most eagerly discussed in marxist economics. This is where Marx's critics thought they had discovered the crucial error in his theory of value; an error which from the outset disqualified it as a scientific theory. And what does this fatal error consist in? Simply in the fact, so the critics think, that Marx was unable to prove his thesis of the reduction of skilled labour to simple average labour, and, driven into an impasse, sought refuge in a circular explanation. Thus, he writes in *Capital* : 'More complex labour counts only as intensified, or rather multiplied simple labour, so that a smaller quantity of complex labour is considered equal to a larger quantity of simple labour.' And immediately after : 'Experience shows that this reduction is constantly being made. A commodity may be the outcome of the most complicated labour, but through its value it is posited as equal to the product of simple labour, hence it represents only a specific quantity of simple labour. The various proportions in which different kinds of labour are reduced to simple labour as their unit of measurement are established by a social process that goes on behind the backs of the producers; these proportions appear to the producers to have been handed down by tradition. In the interests of simplification we shall henceforth view every form of labour-power directly as simple labour-power; by this we shall simply be saving ourselves the trouble of making the reduction.'[1]

Böhm-Bawerk devotes no less than seven pages to this passage in his well-known criticism of Marx. He begins : 'The fact with which we have to deal is that the product of a day's or an hour's skilled labour is more valuable than the product of a day's or an hour's unskilled labour; that, for example, the day's product of a sculptor is equal to the product of five days of a stone-breaker. Now, Marx tells us that things made equal to each other in exchange must contain "a common factor of the same amount" and this common factor must

[1] *Capital* I, p.135 (44).

be labour and working-time. Does he mean labour in general? Marx's first statements [up to page 138 (45)][2] would lead us to suppose so; but it is evident that something is wrong, for the labour of five days is obviously not the "same amount" as the labour of one day. Therefore Marx in the case before us is no longer speaking of labour as such, but of unskilled labour. The common factor must therefore be the possession of an equal amount of labour of a particular kind, viz. unskilled labour. If we look at this dispassionately, however, it fits still worse, for in sculpture there is no "unskilled" labour at all embodied, much less therefore unskilled labour equal to the amount in the five days' labour of the stone-breaker. The plain truth is that the two products embody different kinds of labour in different amounts!'

'Marx certainly says that skilled labour "counts" as multiplied unskilled labour, but "to count as" is not "to be", and the theory deals with the being of things. People may naturally consider one day of a sculptor's work as equal in some respects to five days of a stone-breaker's work, just as they may also consider a deer as equal to five hares. But a statistician might with equal justification maintain with scientific conviction there were 1000 hares in a cover which contained 100 deer and 500 hares, as a statistician of prices or a theorist of value might seriously maintain that in the day's product of a sculptor five days of unskilled labour are embodied, and that this is the true reason why it is considered in exchange to be equal to five days' labour of a stone-breaker.'

This leads on to a long example as to what might be proved, 'if we resorted to the verb "to count" whenever the verb "to be" etc. landed us in difficulties'. However, we can easily dispense with this example for – as we shall soon see – the entire 'substitution', for which Marx was allegedly responsible, is based on mere hair-splitting. However, we now come to Böhm's main argument. Marx – he reminds us – appeals to 'experience' and 'the social process behind the backs of the producers', which supposedly 'proves' the reducibility of skilled labour to unskilled, average labour. However, it is at precisely this point, considers Böhm, 'that we stumble against the very natural, but for the marxian theory very compromising circumstance that the standard of reduction is determined solely by the *actual exchange relations themselves*. But in what proportions skilled labour is to be translated into terms of simple labour in the valuation of their products is not determined nor can it be determined *a priori* by any property inherent in the skilled labour itself. It is

---

[2] That is, up to the section dealing with qualified labour.

rather the actual result alone which decides the actual exchange relations. Marx himself says : "their value makes them equal to the product of unskilled labour", and he refers to a "social process beyond the control of the producers which fixes the proportions in which different kinds of labour are reduced to unskilled labour as their unit of measure", and says that these proportions therefore "seem to be given by tradition".' However, 'under these circumstances what is the meaning of the appeal to "value" and the "social process" as the determining factors of the standard of reduction? Apart from everything else it simply means that Marx is arguing in a complete circle. The real subject of inquiry is the exchange relations of commodities; why for instance a statuette which has cost a sculptor one day's labour should exchange for a cart of stones which has cost a stone-breaker five days' labour, and not for a larger or smaller quantity of stones, in the breaking of which ten or three days' labour have been expended. How does Marx explain this? He says the exchange relation is this, and no other – because one day of sculptor's work is reducible exactly to five days' unskilled work. And why is it reducible to exactly five days? Because experience shows that it is so reduced by a social process. And what is this social process? The same process that has to be explained; that very process by means of which the product of one day of sculptor's labour has been made equal to the value of the product of five days of common labour. But if as a matter of fact it were exchanged regularly against the product of only three days of simple labour Marx would equally bid us accept the rate of reduction of 1 :3 as the one derived from experience and would found upon it and explain by it the assertion that a statuette must be equal in exchange to the product of exactly three days of a stone-breaker's work, not more and not less. In short, it is clear that we shall never learn in this way the actual reasons why products of different kinds of work should be exchanged in this or that proportion. Marx tells us, though in slightly different words, because according to experience they do exchange in this way !' Böhm-Bawerk concludes : 'These are the two ingredients of the marxian recipe . . . : the substitution of "to count" for "to be" and the explanation in a circle which consists in obtaining the standard of reduction from the actually existing social exchange relations which themselves need explanation. In this way Marx has settled his accounts with the factors that most glaringly contradict his theory.'[3]

This then is Böhm-Bawerk's argument, which has been repeated

[3] Böhm-Bawerk, *Karl Marx and the Close of his System* (1896), ed. P.Sweezy, Clifton: Augustus Kelley 1973, pp.81, 82, 83, 86.

so often since then that it now belongs to the 'indispensable fund' of every academic and non-academic critic of Marx.[4] To begin with, we must object to one detail in his argument – namely that he selects a sculptor to be the representative of skilled labour. Such an example only serves to confuse any discussion of Marx's theory of value because Marx, from the outset, excluded 'artistic workers' from the scope of his work, i.e. from his theory of value.[5] Let us therefore leave the sculptor to one side (whether he be a Cellini, as in Böhm's example, or not) and go back to Ricardo's original comparison between a jeweller and a 'common labourer'.

It is clear that their respective products – disregarding the labour objectified in the raw materials and means of labour – 'embody different kinds of labour in different amounts'. However, isn't this also the case if we compare, for example, the labour of a stone-breaker with that of a bricklayer, a car worker or a porter, as any simple, unskilled labour is, depending on its specific properties, different from every other unskilled labour? This is surely not a unique property of skilled labour as such. On the other hand the amount of value-creating labour in the products of the stone-breaker, bricklayer or car worker is in no way known from the outset, even if we know that they have all worked for the same amount of time, for we do not yet know whether they have produced their product under 'socially normal conditions of production' and with the 'socially average level of skill and intensity of labour'. (For example, if the labour of a textile worker from a particular firm is especially productive or intensive, then it might be that the product of half a day of his labour might exchange for an entire day's work by a stone-breaker.) In order for their products to be measured as values, the various labours contained in these products must be reduced to 'undifferentiated uniform human labour'; 'only then can the amount of labour embodied in them be measured according to a common measure, according to time'.[6]

It is inexplicable why these qualitative and quantitative distinctions between the work of different workers only occurred to Böhm-Bawerk when he came to look at skilled labour. Or is this a case of the prejudices of the 'educated classes', according to which

---

[4] One critic who repeats it is Rudolf Schlesinger, author of *Marx, His Time and Ours* (1950). He writes (p.129): 'This problem is certainly the most serious difficulty met by an inherent criticism of marxist economics . . . Should no one succeed in solving the problem', then Marx's theory of value must surely be finally laid to rest.

[5] Cf.*Capital* III, pp.759, 633; and *Theories* I, p.267-68, 410-11.

[6] *Theories* III, p.135.

the labour of certain 'higher professions' (e.g. sculptors) – which on the one hand are not to count as 'unproductive', but on the other should be distinguished from all other labours – are fundamentally distinct from those of the 'common labourer'? So distinct that the latter can be easily reducible to 'undifferentiated uniform human labour', but not the former? Be that as it may, in this respect Böhm proves too much, and consequently too little as well. He fails to notice that according to his formulation the main attack should be directed at Marx's concept of 'undifferentiated human labour' and should not be confined to the special case of skilled labour, to which Marx later applied this concept. This is because one cannot possibly prove an exception to the rule with arguments which destroy the rule itself. Either, the reasons mentioned by Böhm are correct, in which case no labour is reducible to general human labour, and it is also super-fluous to demonstrate this with special reference to skilled labour; or, they are not valid and other reasons must be found for giving a special status to skilled labour.

The same can also be said for Böhm-Bawerk's remaining objec-tions, since the reduction of all labour to 'undifferentiated human labour', which underlies Marx's concept of value, is in no way given from the outset, but takes place by means of a 'social process behind the backs of the producers'; and Marx adds on the subject of this reduction : 'The total labour-power of society, which is manifested in the values of the world of commodities, counts . . . as one homo-geneous mass of human labour-power, although composed of in-numerable individual units.'[7] Therefore why not raise the reproach of a 'substitution' of 'to be' by 'to count' and of the 'circular argu-ment' here, at the source of Marx's concept of value; why reserve these objections for the secondary question of skilled labour?!

2. Thus Böhm's investigation leads us back to the concept of 'undifferentiated' or 'abstract human' labour. What role does this concept play in Marx's theory of value?

At first glance it is obvious that in immediate reality human labour is as diverse as the goods which it produces. 'Let us suppose that one ounce of gold, one ton of iron, one quarter of wheat and twenty yards of silk are exchange-values of equal magnitude . . . But digging gold, mining iron, cultivating wheat and weaving silk are qualitatively different kinds of labour. In fact, what appears objec-tively as diversity of the use-values, appears, when looked at dynamic-ally, as diversities of the activities which produce those use-values.'

---

[7] *Capital* I, p.129 (39).

But this is not all : 'Different use-values are, moreover, products of the activity of different individuals and therefore the result of individually different kinds of labour.'[8] How, then, should labour serve as a common measure of values in the face of the diversity of particular human labours?

This is a problem which was neglected by Ricardo and the other Classical economists, and was first solved by Marx. His analysis of the exchange relation led to the conclusion that as exchange-values, commodities do not contain 'an atom of use-value', and that their value represents 'something purely social'.[9] As exchange-value the economic good is 'no longer a table, a house, yarn or any other useful thing'; but neither 'can it any longer be regarded as the product of the labour of the joiner, the mason or the spinner, or of any other particular kind of productive labour. With the disappearance of the useful character of the products of labour, the useful character of the kinds of labour embodied in them also disappears; this in turn entails the disappearance of the different concrete forms of labour. They can no longer be distinguished, but are all together reduced to the same kind of labour, human labour in the abstract.'[10]

It can be seen : 'Equality in the full sense between different kinds of labour can be arrived at only if we abstract from their real inequality, if we reduce them to the characteristic they have in common, that of being the expenditure of human labour-power, of human labour in the abstract.'[11] At first sight this appears as a purely ideal result; however, in reality this abstraction 'is made every day in the social process of production . . . The conversion of all commodities into labour-time is no greater an abstraction, and is no less real, than the resolution of all organic bodies into air. Labour, thus measured by time, does not seem, indeed, to be the labour of different persons, but on the contrary the different working individuals seem to be mere organs of this labour . . . This abstraction, human labour in general, *exists* in the form of average labour which, in a given society, the average person can perform, productive expenditure of a certain amount of human muscles, nerves, brain etc. It is *simple*

[8] *Contribution*, p.29.

[9] *Capital* I, pp.127, 138, 148, 176-77 (37, 47, 56, 83). Cf. *Theories* III, p.296. 'Use-value expresses the natural relationship between things and men, in fact the existence of things for men. Exchange-value, as the result of the social development which created it, was later superimposed on the word value, which was synonymous with use-value.'

[10] *Capital* I, p.128 (38).

[11] *Capital* I, p.166 (73).

labour, which any average individual can be trained to do and which in one way or another he has to perform.'[12]

And the proof which Böhm-Bawerk so persistently demands? The proof is provided by the capitalist mode of production itself, 'in which individuals can with ease transfer from one labour to another, and where the specific kind is a matter of chance . . . Not only the category, labour, but labour in reality has here become the means of creating wealth in general, and has ceased to be organically linked with particular individuals in any specific form' (as, for example, with the craftsmen of earlier periods). 'Such a state of affairs' (these sentences were written in 1857), 'is at its most developed in the most modern form of existence of bourgeois society – in the United States. Here, then, for the first time, the . . . abstraction of the category "labour", "labour as such", labour pure and simple, becomes true in practice.'[13] Only in the completely developed capitalist mode of production does the entire labour-power of society 'count' (*gelten*), (or let us say 'can be reckoned as' (*zählen*) to avoid the expression forbidden by Böhm as 'one and the same human labour-power'. 'The effect is the same as if the different individuals had amalgamated their labour-time and allocated different portions of the labour-time at their joint disposal to the various use-values. The labour-time of the individual is thus, in fact, the labour-time required by society to produce a particular use-value, that is to satisfy a particular want.'[14] Thus whoever wants to speak of arbitrary abstractions in Marx, should first look at the capitalist production process, where, in fact, labour does not exist for people, but people for labour, and where, in the great majority of cases, this is simply a matter of average labour, at the average pace![15]

This is sufficient elucidation of the basic concept of Marx's theory of value – the concept of 'universal human labour'. We can now

---

[12] *Contribution*, pp.30-31.

[13] *Grundrisse*, pp.104-05. (Cf. the note on p.618 (487) of Volume I of *Capital* on the subject of the ease with which individuals can change their job in the United States.)

[14] *Contribution*, p.32.

[15] Cf. Marx's work of 1847, *The Poverty of Philosophy*, directed against Proudhon : 'If the mere quantity of labour functions as a measure of value regardless of quality, it presupposes that simple labour has become the pivot of industry. It presupposes that labour has been equalised by the subordination of man to the machine or by the extreme division of labour; that men are effaced by their labour; that the pendulum of the clock has become as accurate a measure of the relative activity of two workers as it is of the speed of two locomotives. Therefore we should not say that one man's hour is worth another man's hour, but rather that one man during an hour is worth just as much as

understand why, according to Marx, the values of commodities 'are only social functions of those objects and have nothing to do with their natural qualities',[16] and why, in order to be at all able to measure these values by the labour-time contained in them, we must trace different labours themselves back to undifferentiated, equivalent, simple labour, in which both the individuality of the workers and the concrete character of their activities seem to be extinguished. Naturally this does not mean that value-creating labour is a mere phantom : rather, what underlies it is the very real situation in a producing (i.e. commodity-producing) society, a situation which for its part rests on the no less real 'physiological truth' that any human labour is the 'expenditure of human brain, nerves, muscles and sense-organs'.[17] However, physiological labour is not yet economic labour. This, rather, presupposes the social process of the equalisation of originally different, unequal labours;[18] a process which takes place 'behind the backs of the producers' (in production itself, and therefore also in exchange), and which the category of 'abstract human' labour simply reflects.[19]

3. However, let us return to the real matter at hand – Böhm-Bawerk's objection to Marx's reduction of skilled to unskilled labour. First of all we want to anticipate the somewhat curious objection that there has been a 'substitution'. According to Böhm, Marx was supposed to have 'substituted' what skilled labour 'counts as' for what this labour actually 'is', in order to save having to offer 'proof' of this reduction; whereas everyone knows that scientific theory only deals with the 'being' of things. How little in fact this objection 'counts for'

another man during an hour. Time is everything; man is nothing; he is at the most time's carcass . . . Quantity alone decides everything; hour for hour, day for day; but this equalising of labour is not by any means the work of M. Proudhon's eternal justice; it is purely and simply a fact of modern industry.' (pp.53-54.)

[16] *Wages, Price and Profit. Selected Works*, p.201.

[17] *Capital* I, p.164 (71).

[18] 'On the market products are not exchanged in terms of equal, but of equalised quantities of labour.' (I.I.Rubin, *Essays on Marx's Theory of Value*, p.169.)

[19] It is clear how little Böhm really grasped the basis of Marx's theory of value from his comparison of 'general human' and 'simple average labour', where he designated the latter as a 'particular kind of labour' in a crudely naturalistic way. A fine 'particular kind of labour', 'for which any average individual can be trained' and which the average individual 'must carry out in one form or another'. (This all derives from an elementary confusion; he overlooks that 'simple average labour' can indeed be counterposed to skilled labour as a 'particular kind of labour' – but not to 'general human labour', as it is part of the latter's definition.)

can be best seen by comparing it with two sentences from Marx on the subject of 'skilled labour'. Thus, in the passage in Volume I, quoted by Böhm : 'More complex labour counts only as intensified, or rather, as multiplied simple labour, so that a smaller quantity of complex labour is considered equal to a larger quantity of simple labour.'

How does the parallel passage read in the *Contribution*, published eight years previously? 'But what is the position with regard to more complicated labour . . .? This kind of labour resolves itself into simple labour; it is simple labour raised to a higher power, so that, for example one day of skilled labour may equal three days of simple labour.'[20]

It is clear that both sentences say the same thing : and thus Marx could achieve his purpose without the substitution attributed to him, by saying not 'counts as resolved', but rather 'resolves itself' ! So where is the crucial 'ingredient of his recipe' (to which Böhm-Bawerk – basing himself on one single word – devotes two pages of his critique)?

In fact it is rather disparaging to go into this hair-splitting, which was so unworthy of the leading theoretician of the 'Austrian' school. But couldn't something in fact be learnt from this 'objection'? In fact, why does Marx use the expression 'counts' in this, and several other, passages? Simply to indicate that the value-creating quality of human labour is not a naturally given fact from the outset, but is rather the result of the equalisation of different labours which takes place in a social process. Thus, lurking behind Böhm's curious objection is a naïve naturalistic conception of the labour theory of value, which although having nothing to do with Marx, has a lot to do with his critics' lack of understanding.[21]

Now to the last, and most important of Böhm's arguments – to the famed vicious circle which he discovered in Marx. Is it true that Marx was able to base his theory of the higher value-creating power of skilled labour on nothing more than the workings of the market, where the products of skilled labour receive a higher valuation than those of unskilled labour?

This is yet another example of how fundamentally Böhm misunderstood Marx's theory of value. Here he overlooks the fact that before Marx came to deal with what was for him the secondary question of skilled labour, he had already solved the underlying problem of the reducibility of all labours (be they skilled or unskilled) to 'undifferentiated, uniform, simple labour'; he therefore no longer needed

[20] *Contribution*, p.31.
[21] This also applies to more recent critics of Marx's theory of value such as Schumpeter, Robinson etc.

to 'prove' this already established result once more with respect to skilled labour. (This could only become a problem for someone who regards the work of a skilled worker – such as an engineer or mechanic – as different in principle to all other labour.) Of course this does not mean that the question of skilled labour presents no problem in itself. However, the issue is not whether this kind of labour is reducible in principle to unskilled labour, whether it does in fact represent a simple multiple of unskilled labour; but rather by what laws does this reduction take place and how is this multiple to be measured. And these are naturally two quite different questions! Even Böhm would have noticed this if he had looked at the *Contribution* as well as *Capital* (in fact Marx himself refers to the *Contribution* as a necessary complement to the theoretical chapters on money and value).[22] Thus, in this text, it states on the question of skilled labour; 'The laws governing this reduction [of skilled to unskilled labour] do not concern us here. It is however clear that the reduction is made, for as exchange-value, the product of highly skilled labour is equivalent, in definite proportions, to the product of simple average labour; thus being equated to a certain amount of this simple labour.'[23]

We consider that this passage alone should put a stop to all talk of Marx's alleged 'circular explanation', since if, according to Marx, the higher value-creating power of skilled labour is simply deducible from the higher valuation of the products it produces on the market, why did he speak, in the same breath, of the particular laws which govern the reduction of skilled to unskilled labour?[24] How is this compatible with Böhm's assertion that according to Marx 'the standard of reduction is determined by nothing other than the actual exchange relations themselves'? It is not compatible at all. It is hardly surprising that Böhm took no notice of this passage, which is so unfavourable for his interpretation. This may indeed have served as a polemic against Marx; but hardly as a scientific investigation.

II. MARX'S PROBABLE SOLUTION

Marx accepted Ricardo's assertion that the processes on the market for commodities themselves confirm that a reduction of skilled to unskilled labour takes place. (It is difficult to see why Böhm did not

[22] *Capital* I, p.89 (75).
[23] *Contribution*, p.31.
[24] In the section quoted by Böhm, Marx in fact speaks of the 'effort of reduction' which he wishes, initially, to avoid.

refer to Ricardo's authorship of this argument in his criticism of Marx.) However, Marx already notes in his *Exzerpten* of 1851 : 'Ricardo provides no further development of this.'[25] However, the fact that he considered[26] working out such a 'development' himself (probably in the planned *Book on Wage-Labour*) can be seen in a passage from the *Theories* directed against Bailey. 'His last objection' (to Ricardo's theory of value) 'is this : The value of commodities cannot be measured by labour-time, if the labour-time in one trade is not the same as in the others, so that the commodity in which, for example, 12 hours of an engineer's labour is embodied has perhaps twice the value of the commodity in which 12 hours of the labour of an agricutural labourer is embodied. What this amounts to is the following : A simple working day, for example, is not a measure of value if there are other working days which, compared with days of simple labour, have the effect of composite working days.[27] Ricardo showed that this fact does not prevent the measurement of commodities by labour-time if the relation between unskilled and skilled labour is given. He has indeed not described how this relation develops and is determined. This belongs to the definition of *wages*, and, in the last analysis' – now the surprise – 'can be reduced to the *different values of labour-power itself*, that is, its varying production costs (determined by labour-time).'[28]

How should we interpret this interesting passage? At any rate not in the sense that the thesis, according to which any complex labour simply represents multiplied simple labour, is first to be 'proved'; this task was already carried out in the theory of value, by the reduction of all labour to simple average labour. The question is, therefore, not

[25] *Grundrisse*, German edn. p.787.

[26] Cf. *Grundrisse*, p.846. 'Of course, labour is distinct qualitatively as well, not only insofar as it is performed in different branches of production, but also more or less intensive etc. The way in which the equalisation of these differences takes place, and all labour is reduced to unskilled simple labour, cannot of course be examined yet at this point. Suffice it that this reduction is in fact accomplished with the positing of products of all kinds of labour as values. As values, they are equivalents in certain proportions; the higher kinds of labour are themselves appraised in simple labour. This becomes clear at once if one considers that e.g. Californian gold is a product of simple labour. Nevertheless, every sort of labour is paid with it. Hence the qualitative difference is suspended, and the product of a higher sort of labour is in fact reduced to an amount of simple labour. Hence these computations of the different qualities of labour are completely a matter of indifference here, and do not violate the principle.'

[27] Hence, Bailey anticipates Böhm-Bawerk's argument here. (Böhm also fails to mention Bailey's polemic against Ricardo in his own critique.)

[28] *Theories* III, p.165. (Cf. the 'Appendix' to Chapter 2 above.)

whether skilled labour is fundamentally capable of reduction to unskilled, but rather, by what standard this occurs, and how the respective labours can be compared with one another. And here Marx, the strict labour-theorist, is 'unorthodox' enough to propose the 'different values of labour-power itself', i.e. basically, the differing training costs of skilled and unskilled labour, as the standard of comparison! A solution which occurred to many – more or less 'orthodox' marxists (to name only Kautsky, C.Schmidt, Bernstein, Hilferding, H.Deutsch, O.Bauer, L.Boudin, Bogdanov, Posniakov, and Rubin) but from which they, for the most part, and with a correct instinct, recoiled, since this solution – from the standpoint of the artificial problem raised by Böhm – would lead without question to the derivation of the values of commodities from the value of labour-power, i.e. would contradict the essence of Marx's theory of value itself. This explains several – in part very ingenious – attempts to fill in the supposed gaps in Marx's theory of value, and in this way ward off Böhm-Bawerk's attack.

We have no intention of increasing the number of these attempts; firstly, because we do not want to measure our abilities against those of the theorists mentioned above; and secondly, because we regard the problem – as they raised it[29] – as non-existent.

Let us stress once more that the question is no longer whether skilled labour represents a simple multiple of unskilled, but simply how this multiple is to be measured. And it is incomprehensible why this should not occur in the way that Marx proposed in the *Theories*. Let us for a moment imagine a socialist society. Even this society, at the outset, will certainly have to deal with the fact of differently qualified labour. Thus here too the question of the reduction of skilled to unskilled labour will be of theoretical – and, above all, practical – significance. It will be significant in two respects; one, as far as the 'reward' of the labour-powers of different qualifications is concerned; and two, because a socialist society will have to make a careful calculation of the labour-power at its disposal and distribute it to the various branches of production.

As far as rewarding skilled labour is concerned, the socialist society, as Engels pointed out, will seek, above all, to equate the 'wages' of skilled workers to those of unskilled. The reason is easy to understand : 'In a society of private producers, private individuals or their families pay the costs of training the qualified worker; hence the higher price paid for qualified labour-power accrues first of all to

---

[29] It should be remembered that at that time no consideration had been given to the structure or the changes in the structure of *Capital*.

private individuals : the skilful slave is sold for a higher price and the skilful wage-earner is paid higher wages. In a socialistically organised society these costs are borne by society and to it, therefore, belong the fruits, the greater "values"[30] produced by compound labour.'[31] (N.B. assuming that this society is capable of bearing the entire costs of training, which at the outset will not necessarily be the case.)

However, what is more important is the second question : how does a socialist society deal with the fact of differently qualified labour in its economic planning? Since the higher power of skilled labour does not arise from any mysterious property possessed by this labour itself, or by its human bearer, it is evident that this can only be based on the empirically given and empirically measurable difference in the training costs of skilled and unskilled workers themselves. Assume that 100 workers who work 10 days are necessary for the completion of a particular project, of which however 10 must be equipped with particular, above-average qualifications, specially for this project. In order to train these workers society must incur certain expenses which, let us say, amount to 200 working days. It is clear then, that these 200 working days must also be 'accounted for' by society, if its economic plans are to have a sound basis. It would therefore allow not 1,000 working days, but rather 1,200 for the carrying out of the project. Thus the distinction between skilled and unskilled labour will in the final analysis be reduced to the difference in the period of training of the various kinds of labour.

The same would apply, *mutatis mutandis*, for the capitalist mode of production as well, except that here there is no central planning organ which would balance the training time of the various kinds of labour, and must, rather, give this task over to the spontaneous forces of the market (the market for labour, and the market for commodities); and furthermore, here, the connection between the training time of different workers must assume the form of interaction between the values of the labour-powers and the values of the commodities which they produce. In this sense Marx states in *Capital* : 'All labour of a higher or more complicated character than average labour is expenditure of labour-power of a more costly kind, labour-power whose production has cost more time and labour, than unskilled or simple labour-power, and which therefore has a higher value. This power being of higher value, it expresses itself in labour of a higher

---

[30] We have placed the word 'values' in inverted commas as it is clearly only used in an analogous sense and could easily lead to misunderstandings.
[31] Engels, *Anti-Dühring*, pp.239-40.

sort, and therefore becomes objectified, during an equal amount of time in proportionally higher values.'[32]

This in no way means that Marx, in contradiction to his theory of value, here derives the value of commodities from the 'value of labour', but rather that, in the social process of the equalisation of different labours, the extra expenditure of labour which capitalist society must employ for the training of skilled workers cannot be *expressed* other than through a higher 'valuation' of the products produced by these labour-powers. (If this were not the case no employer would be prepared to pay skilled workers correspondingly higher wages. The consequence would be a movement of workers out of these trades which would continue until the demand for the products concerned drove up their price, and thereby compelled the training of new skilled workers.)

So much, then, on the solution to the problem of skilled labour as it is touched on in the *Theories*. However the main point was not so much the solution itself but rather the proof that the distinction between skilled and unskilled labour offers no kind of obstruction, in principle, to the explanation of economic phenomena from the standpoint of Marx's theory of value, as was maintained by Böhm-Bawerk and other critics of Marx's theory of value who succeeded him.[33] Indeed, the concept of 'value-creating labour' is not, as Marx remarked in an earlier passage directed against Smith, 'Scottish', i.e. to be taken in a crude naturalistic way, for this reason : 'When we

---

[32] *Capital* I, p.305 (197). [Rosdolsky adds here] We cite the 3rd edition of Volume I here as the last sentence reads somewhat differently in the 4th edition, and because the differing style of the 3rd edition led to a splendid polemic between Hilferding and Bernstein which was first brought to light by the editor of the English translation of Hilferding's work *Böhm-Bawerk's Criticism of Marx* (*op. cit.*), originally published in 1920. Bernstein, taking the passage from the 3rd edition as his basis, maintained that Marx here derives the value of the product from the 'value of labour' (*Die Neue Zeit*, 23 December 1899). Hilferding angrily replied that the passage in question states 'the precise opposite of that which Bernstein wants to deduce from it' and that 'if Bernstein was right he would have had to use the word "*daher*" ("consequently" or "therefore")', which he did not in fact do. But as bad luck would have it, neither Hilferding nor Bernstein took note of the 4th edition of Volume I, where the sentence under discussion was changed by Engels as follows : 'This power being of a higher value, it is therefore expressed in a higher labour, and is consequently objectified in proportionately higher values.' (*Capital* I, p.305 (197). Thus Engels had already introduced the tabooed word '*daher*' into the text ten years previously in 1890, and so gave some assistance to Bernstein ! This merely illustrates what pedantic squabbling over quotations was carried on by the 'orthodox' marxists.

[33] All the less as the differential between the wages of skilled and unskilled work is often simply based on convention. Cf. *Capital* I, p.305 (197).

speak of the commodity as a materialisation of labour – in the sense of its exchange-value – this itself is only an imaginary, that is to say a purely social mode of existence of the commodity which has nothing to do with its corporeal reality; it is conceived as a definite quantity of social labour or of money !'[34]

However, anyone who, like Böhm, demands that Marx's theory of value should deduce the reducibility of skilled labour to unskilled '*a priori*, from some characteristic inherent in skilled labour', merely demonstrates how fundamentally he must have misunderstood this theory.

[34] *Theories* I, p.170.

32.
A Note on the Question of 'False Rationalisation'

According to Otto Bauer (who coined the concept), 'false rationalisation' is rationalisation which, although decreasing the costs of
production in one single enterprise, simultaneously increases the overall social costs of production thus 'making the individual richer and
society as a whole poorer'.[1] 'False rationalisation' is therefore a
phenomenon which is characteristic of the capitalist economic system.
In the capitalist economy labour-power is a commodity whose value,
like that of every other commodity, is determined by the social labour
necessary for its reproduction. The bearer of labour-power, the
worker, nevertheless expends energy in the process of living, as well
as in working, and the former of course even when he or she is unemployed i.e. falls into the ranks of the industrial reserve army. The
worker's labour-power must also continue to be maintained as a
potential object of exploitation by capital during such periods of unemployment and society must therefore provide the unemployed
worker with unemployment pay, 'be this by means of unemployment
insurance, public welfare, or private charity . . . which must be sufficient to replace the energies expended in the process of living'. If an
expenditure of x Marks is needed for the reproduction of the energies
used in the process of living, and y Marks for those in working, then
unemployment pay must amount to at least x Marks, and the wage to
at least x + y Marks, if the worker is to remain capable of working.'
However, the costs of the reproduction of the energies expended in
the workers' life-process are only borne by the employer if the worker
is in his employ : if the worker is made redundant they fall onto
society as a whole. Hence the costs of maintaining unemployed
workers do not constitute 'an element of the production costs of the
individual enterprise, but rather enter into the social costs of production'.
  This distinction is noticeable whenever capitalist rationalisation

[1] Otto Bauer, *Kapitalismus und Sozialismus nach dem Weltkrieg* Vol.I,
Rationalisierung – Fehlrationalisierung, 1931, pp.170-77.

takes place. Because the individual employer is no longer concerned about the workers he has thrown onto the streets and no longer has to provide for their subsistence, it is in his interest continually to 'release labour' by means of technical innovation, the introduction of new machinery etc. (in short – through rationalisation), as long as 'the additional expense on fixed costs which the rationalisation requires, is less than the saving in wages which they bring about'. Hence it is beneficial for the capitalist, as Otto Bauer showed, 'to take rationalisation up to the point where the marginal increase in fixed costs is exactly equal to the savings in wages which this brings about.' In order to illustrate this thesis Bauer cites an example from a report of the committee enquiring into the German economy in 1930. It reads : 'We have completely modernised a small foundry which was equipped with three blast furnaces and reduced the number of workers in one section from 120 to 10 by means of this restructuring, i.e. saved 110 workers. Since workers cost around 4,000 Marks per year at the moment, we have therefore saved 440,000 Marks.[2] The rebuilding cost 2.8 m. Marks, of which 15 % (420,000) are absorbed from our savings in the form of interest and depreciation.' The actual saving was, therefore, a mere 20,000 Marks per annum !

'From the standpoint of capitalist accountancy', says Bauer, 'this rationalisation was still justified'. However, the matter appears quite different when looked at from the standpoint of 'social accounting' : 'If the workers made unemployed by the change remained unemployed for a long period of time, or if they eventually had to move elsewhere to find work, then the additional social cost which the subsistence or moving of the unemployed required was no doubt much greater than the saving of 20,000 Marks.' Bauer therefore considers that for 'society' this changeover signified 'false rationalisation', since, 'from the standpoint of social accounting, technical change will only be worthwhile if it lowers total social costs – i.e. if the saving in capitalist costs is greater than the social expenses required for the maintenance, retraining and relocation of workers made unemployed by technical change.'[3]

Bauer provides a forceful criticism of capitalist rationalisation, which in many instances achieves an increase in profit for the individual employer at the expense of the capitalist economy as a whole (alias, 'society'), and can consequently quite justifiably be characterised, from the standpoint of 'social accounting', as 'thoughtless', 'rash'

[2] As is normal in capitalist language, workers are here equated with tools and raw materials.

[3] Bauer, *op. cit.* pp.169-75.

and 'misconceived', negative rationalisation. For this reason, the concept of 'false rationalisation' certainly has its uses (setting aside the questionable abstraction, 'society').[4] The question is simply, what are the limits within which this concept can be employed? And in what kind of society can one speak of 'false rationalisation' in the above sense? This is a point at which we have to differ from Bauer.

Let us imagine a society in which all the branches of production are collected together in a capitalist trust; that is, where there are no longer individual employers, but rather 'where the bourgeoisie administers the entire economy through its state'. In such a society social accounting (i.e. for capital as a whole) would in fact have to replace private business accounting. A society of this kind would therefore only rationalise if the savings in costs of 'living labour', in wages, were not outweighed by increased expenditure for the maintenance of the army of the unemployed (or would at most equal it). Thus, for this society the limits of rationalisation would be more narrowly drawn than for private capitalism – they would merely be able to rationalise more carefully and prudently (although perhaps also with more continuity). (And we would add that the concept of 'false rationalisation' only has a meaningful application in such a society i.e. as an economic measure which is miscalculated, negative in its consequences and which would burden the capitalist economy with the superfluous costs of the maintenance of labour-power, and which could therefore be criticised from the viewpoint of the 'general interest'.)

---

[4] 'Nothing is more erroneous than the manner in which economists as well as socialists' (Marx refers to Proudhon here) 'regard society in relation to economic conditions . . . This so-called contemplation from the standpoint of society means nothing more than the overlooking of the differences which express the social relation (relation of bourgeois society). Society does not consist of individuals, but expresses the sum of interrelations, the relations within which these individuals stand. As if someone were to say: Seen from the perspective of society, there are no slaves and no citizens: both are human beings. Rather, they are that outside society. To be a slave, to be a citizen are social characteristics, relations between human beings A and B. Human being A, as such, is not a slave. He is a slave in and through society.' (*Grundrisse*, pp.264-65.) Cf. *Capital* II, p.437: 'Speaking of the point of view of society . . . we must not lapse into the manner copied by Proudhon from bourgeois economy and look upon this matter as though a society with a capitalist mode of production, if viewed en bloc, as a totality, would lose this its specific historical and economic character. No, on the contrary. We have, in that case, to deal with the aggregate capitalist.' Hence Marx's usual expression: 'from the standpoint of society or the capitalist class'. (See *Theories* II, p.415, *Capital* II, p.337.) But what is the distinction between Bauer's 'social production accounting', taken from American economists (e.g. J.M.Clark), and Proudhon's 'manner'?

This is how the question appears from the standpoint of state capitalism. But how would it appear in a socialist society, where rationalisation would be tied up with a reduction in general working time, and in which there would no longer be a problem of unemployment (and therefore no problem of the consequent costs of re-training or re-locating the unemployed)? Since wage-labour would no longer exist, this society would clearly have to take account not of the costs of labour-power (as in capitalism), but of the amount of labour itself in its 'cost-accounting'. Consequently, even those technical changes which appear to be misplaced from the standpoint of the state or private capitalist economy, would prove tenable here. 'False rationalisation' would only occur if the new machines etc. had cost society more than (or the same as) they saved on labour (not on the payment for labour-power!). Therefore the limits of rationalisation would be much wider than in the capitalist economy; socialist society could 'rationalise' much more rapidly and on a more generous scale, could develop the productive powers of social labour in a much more dynamic way!

Strange as it may seem, Otto Bauer came to a diametrically opposed view: He writes, in conclusion: 'The source of this false rationalisation would only be prevented in a society in which the factories belong to the state and the very same state had to carry at the same time, the costs of unemployment pay, and the costs of retraining and relocation of workers. Instead of capitalist accounting there would be social accounting. The state would then only have an interest in rationalisation if the saving in the production costs in the individual plant were larger than the costs of the unemployment pay of the workers temporarily dispensed with by the rationalisation measures. This state would naturally also carry out rationalisation. But it would only rationalise to the extent that it could transfer the workers made superfluous by rationalisation in one plant to other plants or branches of production. Rationalisation would not proceed by leaps and bounds, as in the period 1924-29, but would occur more slowly, but steadily and continuously. Carried out only at the speed at which workers who are no longer required could be transferred to other branches of production, rationalisation in a socialist society would no longer be obtained at the expense of long-term mass unemployment.'[5]

It is clear, though, that those features and benefits which Bauer denotes as being characteristic of a socialist society, in fact apply to a

[5] Otto Bauer, *op. cit.* pp.179-80.

The question of 'false rationalisation' · 525

state capitalist system.[6] Not only are we still confronted with the problem of unemployment, but Bauer's 'socialist' society still bases its 'accounting' on the cost of labour-power ('capitalist costs'), rather than on the labour itself which the products require. But this is the very point which serves to distinguish a socialist society from a capitalist one!

We read in Volume I of *Capital*: 'The use of machinery for the exclusive purpose of cheapening the product is limited by the requirement that less labour must be expended in producing the machinery than is displaced by the employment of that machinery. For the capitalist, however, there is a further limit on its use. Instead of *paying for the labour*, he pays only the value of the labour-power employed; the limit to his using a machine is therefore fixed by the *difference between the value of the machine and the value of the labour-power replaced by it*. Since the division of the day's work into necessary labour and surplus-labour differs in different countries, and even in the same country at different periods, or in different branches of industry; and further, since the actual wage of the worker sometimes sinks below the value of his labour-power, and sometimes rises above it, it is possible for the *difference between the price of the machinery and the price of the labour-power replaced by that machinery to undergo great variations while the difference between the quantity of labour needed to produce the machine and the total quantity of labour replaced by it remains constant*. But it is only the former difference that determines the cost to the capitalist of producing a commodity, and influences his actions through the pressure of

[6] The term 'state capitalism' refers here merely to a developmental tendency, and not to an actually existing form of capitalism. Even if such a form were to come into existence, this would in no way signify the end of capitalism, for several capitals, organised by the state, would still confront one another. (Cf. Note 117 on p.42 above.) Cf. also Trotsky's arguments: 'Theoretically, to be sure, it is possible to conceive a situation in which the bourgeoisie as a whole constitutes itself a stock company which, by means of its state, administers the whole national economy. The economic laws of such a regime would present no mysteries. A single capitalist, as is well known, receives in the form of profit, not that part of the surplus-value which is directly created by the workers of his own enterprise, but a share of the combined surplus-value created throughout the country proportionate to the amount of his own capital. Under an integral "state capitalism", this law of the equal rate of profit would be realised, not by devious routes – that is, competition among different capitals – but immediately and directly through state book-keeping. Such a regime never existed, however, and, because of profound contradictions among the proprietors themselves, never will exist – the more so since, in its quality of universal repository of capitalist property, the state would be too tempting an object for social revolution.' (*The Revolution Betrayed*, pp.245-46.)

competition.' (Marx adds in a footnote : 'The field of application for machinery would therefore be entirely different in a communist society from what it is in a bourgeois society.'[7])

This remark illustrates clearly enough the difference between the 'cost accounting' of a capitalist and a socialist society. However, in Volume I of *Capital* this distinction is only hinted at, whereas a detailed discussion can be found in Volume III.[8] Here, in a passage edited by Engels, we read : 'The value of a commodity is determined by the total labour-time of past and living labour incorporated in it. The increase in labour productivity consists precisely in that the share of living labour is reduced while that of past labour is increased, but in such a way that the total quantity of labour incorporated in that commodity declines; in such a way, therefore, that living labour decreases more than past labour increases. The past labour contained in the value of a commodity – the constant part of capital – consists partly of the wear and tear of fixed, partly of circulating constant capital entirely consumed by that commodity, such as raw and auxiliary materials. The portion of value deriving from raw and auxiliary materials must decrease with the [increase of the] productivity of labour, because with regard to these materials the productivity expresses itself precisely by reducing their value. On the other hand, it is most characteristic of rising labour productivity that the fixed part of constant capital is strongly augmented, and with it that portion of its value which is transferred by wear and tear to the commodities. For a new method of production to represent a real increase in productivity, it must transfer a smaller additional portion of the value of fixed capital to each unit of the commodity in wear and tear than the portion of value deducted from it through the saving in living labour; in short it must reduce the value of the commodity . . . This reduction of the total quantity of labour going into a commodity seems, accordingly, to be the essential criterion of increased productivity of labour, no matter under what social conditions production is carried on. Productivity of labour, indeed, would always be measured by this standard in a society in which producers regulate their production according to a preconceived plan, or even under simple commodity production. But how does the matter stand under *capitalist* production?'

At this point Engels introduces the following example : 'Suppose a certain line of capitalist industry produces a normal unit of its

---

[7] *Capital* I, p.515 (392).
[8] Cf., however, *Grundrisse*, pp.776-77, 819-20.

commodity under the following conditions :[9] The wear and tear of fixed capital amounts to $\frac{1}{2}$ shilling per piece; raw and auxiliary materials go into it to the amount of $17\frac{1}{2}$ shillings per piece; wages, 2 shillings; and surplus-value, 2 shillings at a rate of surplus-value of 100%. Total value = 22 shillings . . . The cost-price of the commodity = $\frac{1}{2} + 17\frac{1}{2} + 2 = 20$s, the average rate of profit $2/20 = 10\%$, and the price of production per piece of the commodity, like its value = 22s. Suppose a a machine is invented which reduces by half the living labour required per piece of the commodity, but trebles that portion of its value accounted for by the wear and tear of the fixed capital. In that case, the calculation is : Wear and tear = $1\frac{1}{2}$ shillings, raw and auxiliary materials, as before, $17\frac{1}{2}$s., wages, 1s., surplus-value 1s., total 21s. The commodity then falls one shilling in value; the new machine has certainly increased the productivity of labour. But the capitalist sees the matter as follows : his cost price is now $1\frac{1}{2}$s. for wear, $17\frac{1}{2}$s. for raw and auxiliary materials, 1s. for wages, total 20s. as before. Since the rate of profit is not immediately altered by the new machine, he will receive 10% over his cost price, that is 2s. The price of production, then, remains unaltered = 22s., but is 1s. above value. For a society producing under capitalist conditions of production the commodity has not cheapened. The new machine is no improvement for it. The capitalist is, therefore, not interested in introducing it. And since its introduction would make his present, not as yet worn-out, machinery simply worthless, would turn it into scrap-iron, hence would cause a positive loss, he takes good care not to commit what would be for him a utopian mistake.' Engels concludes : 'The law of increased productivity of labour is not, therefore, absolutely valid for capital. So far as capital is concerned, productiveness does not increase through a saving in living labour in general, but only through a saving in the *paid* portion of living labour, as compared to labour expended in the past . . . Here the capitalist mode of production is beset with another contradiction. Its historical mission is unconstrained development in geometrical progression of the productivity of human labour. It goes back on its mission whenever, as here, it checks the development of productivity. It thus demonstrates again that it is becoming senile and that it is more and more outlived.'[10]

We considered that this long quotation was necessary, as it expands on and complements Marx's remarks in Volume I of *Capital*

---

[9] Under the assumption that 'the capital in this line of production has the average composition of social capital'.

[10] *Capital* III, pp.260-62.

S

in a particularly precise and illuminating way. One thing is clear : this solution follows necessarily from marxist economy theory. Otto Bauer too, knew this only too well in his time, as his earlier writings show. Thus we read in his first work : 'Capitalist production not only reduces the full use of available human labour-power, it also hinders the application of the most effective methods of production. . . . Socialist society will be able to employ a machine, if it saves more labour than is required for its manufacture; by contrast, the capitalist mode of production is only able to use a machine if it saves more wages than it costs. The lower the wages, the more difficult it is to introduce new machines, or to make use of technical progress. Since the wage can only be the form of appearance of the value of labour-power and never of the product of labour, capitalist society could never employ all those machines which a socialist society could put at its service. But there is still more !'

In addition we have the effects of the equalisation of the individual rates of profit so as to form a general rate of profit : 'The production price of the machine is permanently higher than its value . . . because this contains a part of the surplus-value produced in other branches of production and appropriated by the manufacturers of iron and machines by virtue of the size of their physical capital . . . We can then add, that the higher production price of the machine which is always more than its value . . . is a further limitation to the replacement of manual labour by more productive machinery. And finally there is one additional reason ! The cartels and trusts in the coal and iron industries increase the price of their products – of machines – still further above the production price created by free competition, i.e. make machine production even more expensive, and are therefore a further limitation to technical progress. The socialist mode of production would do away with all these limits at one blow : for such a society any machine can be used which saves more labour than it itself requires.'[11]

This shows how well Otto Bauer could write in his earlier years ! Not a word about the quasi 'social accounting', working with the amount required for wages; and also no warnings about a careful, gradual, cautious pace of rationalisation under socialism ! How does one then explain the fact that Bauer later came to diametrically opposed conclusions – despite the fact that he belonged to the marxist school ?

In fact the explanation is very simple. Twenty-five years elapsed

[11] *Die Nationalitätenfrage und die Sozialdemokratie*, 2nd edn., pp.97-98.

and Saul became Paul. Once he had been converted to reformism Otto Bauer had to look at the question of the socialist transformation of society through the eyes of a reformist 'realpolitiker', and 'statesman'. This is how his miraculous idea arose of a socialist society, in which capitalist methods of accounting were still applicable; a society which numbered among its normal institutions, offices for paying unemployment benefit. The 'socialist society' which he speaks of in 1932 is nothing more than a state capitalist society; a society which has negated capitalism simply in the sense that it has transferred the control of the means of production to the state, without introducing the socialist mode of production, without making the producers themselves the real managers of the economy. Thus the criticism which can be levelled at Otto Bauer is that he neglected the most important thing about this society, which was at that time hypothetical, but which is sought in practice today by reformism; that he limited himself merely to the problem of the transformation by the state of 'organised capitalism' so that finally the decisive distinction between socialism and capitalism, in relation to the development of the forces of production, disappeared. Was this a merely theoretical error? By no means. A highly characteristic concept of socialism hides behind the academic question as to the optimal speed of rationalisation under socialism, a concept which basically equates socialism with state capitalism. This could perhaps be overlooked when Otto Bauer wrote his book; however, today, after the crucial experiences of the last decades, the question of the distinction and antithesis between the socialist and state capitalist economic order must be recognised as one of the central questions of the workers' movement, since it seems certain that this contrast will play a prominent role in the future struggles of the working class and in the future ideological disputes within the socialist camp. Seen in this light Otto Bauer's error takes on a completely different appearance.

33.
Joan Robinson's Critique of Marx

It has often been stated that the differences between the two opposed schools of political economy – the 'academic' and the marxist – have grown so large that the adherents of one school can scarcely understand the language of the other. A striking example of this gulf is provided by Joan Robinson. This distinguished economist was anxious to do justice to the 'rough and gloomy grandeur'[1] of Marx's theoretical system; in fact she succeeded in doing nothing more than producing a further caricature of marxism. The reason for this does not of course lie in some kind of individual failing (Joan Robinson is an excellent economist); it lies deeper than this, in the method with which she approached her task. It is obvious that a critic of Marx who admits to having little regard for Marx's method, and who dismisses it as 'Hegelian stuff and nonsense'[2] must necessarily misunderstand and misinterpret even its most elementary principles. She might, perhaps, understand what Marx said literally, but never 'what he really meant'.

I. MARX'S THEORY OF VALUE

Naturally, the main target of Robinson's criticism is Marx's theory of value, since if she succeeds in demolishing this cornerstone of Marx's system, the basic assumptions of so-called academic theory can remain intact and be used to yield some kind of quasi-socialist conclusions. The result would be the creation of a neatly pruned, Fabianised and Keynesianised Marx.

1. Marx as a 'value fetishist'

Let us start with a few remarks on Robinson's critical method. In common with so many of her predecessors she divided Marx into

[1] Joan Robinson, *An Essay on Marxian Economics*, p.2.
[2] Robinson, *On Re-reading Marx*, p.20.

two different, in fact opposed, beings; the 'Hegelian metaphysician' of Volume I of *Capital* and the Marx who was governed by common sense in Volume III. What the latter wrote, could, to some extent, be reconciled with reality (especially if corrected from the standpoint of 'modern economics'). This is because, basically, the theory of value as set out in Volume III is 'everyone's theory' – in contrast to the theory of 'absolute value' in Volume I[3] which is 'pure dogmatism' and therefore simply 'indigestible'.[4] Let us therefore begin with this 'indigestible' part.

Robinson writes : 'Ricardo sought to find in labour cost a measure of value which would be invariable in the same manner as a measure of length or of weight, and Marx, though he did not read Ricardo's essay, *Absolute Value and Exchangeable Value,* echoes the same thought when he looks for the "something" in common between commodities of equal exchange-value, which "cannot be a geometrical, physical, chemical or other natural property of commodities".'[5]

He regarded value 'as a characteristic inherent to these commodities, as weight or colour'.[6] This concept of value, continues Robinson, 'is essentially pre-marxist', and is in glaring contradiction to the fundamental principles of Marx's doctrine. 'For one of Marx's greatest contributions to analysis is the distinction between the "forces of production" and the "relations of production"; that is, between the technical relations of man with his physical environment, and the economic relations of a man with his neighbours in society; and the notion of the fetishism which attaches itself to exchangeable commodities[7] – qualities arising out of relations between people appearing as relations between things.'

But, 'it takes the author of an original idea a long time to see all its implications – there are many examples of pre-Keynesian thought in the *General Theory*'! Hardly surprising then that Marx overlooked the simple fact that, 'weight and length are technical, value is social. Robinson Crusoe provides a touchstone for the distinction;

---

[3] Moreover, Robinson confuses Marx with Ricardo here. Marx never used the expression 'absolute value', and in fact rejected it, because it implies an independence of value from social relations. (See *Theories* III, pp.130-31, 134.)

[4] Robinson, 'The Labour Theory of Value : A Discussion', in *Science and Society*, 1954.

[5] *ibid.*

[6] Robinson, *Collected Economic Papers*, p.147.

[7] One of these words is clearly superfluous, as every 'commodity' is 'exchangeable' and every 'exchangeable good' is a 'commodity'.

s*

weight and length had the same meaning for him on his island as at home : purchasing power had no meaning at all.'[8] The task facing Marx's critics is then one of, at long last, freeing Marx's theory from this terrible inconsistency.

This then is Robinson's view. Her entire argument can be reduced to two simple statements : (1) to say that value is something inherent in the commodity is to regard value as a technical category, and (2) to say that labour is the essence of value is to see in the cost of labour 'the invariable measure of value'.

Both assertions are based on trivial misunderstandings. As we have just seen, Marx asserted that the 'common element' which determined the value of commodities 'could not be a geometrical, physical, chemical, or other natural property of commodities'.[9] But what else can it be? Is it their common social nature which we repeatedly read about in Marx? No, explains Joan Robinson : this is where you make your mistake ! Since what is 'common' to commodities must be inherent in them; and further, what is 'in them' can only be of a material, and not of a social nature . . . Thus, only two solutions are possible : either value is a social phenomenon, in which case it cannot simultaneously be an 'inherent' characteristic of commodities; or it is in fact 'inherent' in them, in which case it must be characterised as 'analogous to weight and colour', i.e. as a natural attribute. Isn't it therefore obvious that Marx simply confused value (which clearly represents a social relation) with a natural or technical category, and thus himself became a victim of the 'commodity fetishism' described so admirably in his book?

Of course, nothing is more agreeable than to see an academic economist standing up so energetically for the exclusive social character of the concept of value. (How this is compatible with the dominant role of 'utility' in modern economics is quite another question.) However, is it not rather foolish to raise this in relation to Marx, who was the first to recognise the pre-eminently social character of value, and who made this the cornerstone of his entire system?

He writes in his principal work : 'However, let us remember that commodities possess an objective character as values only insofar as they are all expressions of an identical social substance, human labour.'[10] 'As crystals of this social substance, which is common to them all, they are values – commodity-values.'[11] For, 'the commodity-

---

[8] *Science and Society, op. cit.*
[9] *Capital* I, p.127 (37).
[10] *ibid.* p.138 (47).
[11] *ibid.* p.128 (38).

form, and the value-relation of the products of labour' have 'absolutely no connection with the physical nature of the commodity and
the material relations arising out of this. It is nothing but the definite
social relation between men themselves which assumes here, for them,
the fantastic form of a relation between things.'[12]

And Marx adds, as if anticipating the kind of criticism his theory
might encounter in the future : 'Just as the body of the iron as a
measure of weight represents weight alone, in relation to the sugar-
loaf, so, in our expression of value, the body of the coat represents
value alone. Here, however, the analogy ceases. In the expression of
the weight of the sugar-loaf, the iron represents a natural property
common to both bodies, their weight; but in the expression of value
of the linen the coat represents a supra-natural property : their value,
which is something purely social.'[13]

Marx makes the distinction between the weight-relation of the
two bodies and the value-relation of two commodities absolutely clear
here; the former is a material relation, the latter a purely social one.
However, this did not prevent his Keynesian critic from imputing
exactly the opposite view to him : the concept of value as 'a quality
analogous to weight or colour' and, on top of that, to lecture him on
the distinction between the 'technical' and the 'social' – two concepts
which must have been clear even to Robinson Crusoe, although the
poor man was never a professor of economics. But how could Joan
Robinson arrive at such grotesque conclusions? The explanation
clearly has to be looked for in the field of methodology.

Marx asks, how can we designate labour as the substance of
value if, in actual fact, each concrete labour serves a different aim,
and is performed by different individuals of differing ability, skill
etc.? How can the infinite variety of the different kinds of professional
and individual labour be reduced to a common denominator? His
answer is that it is possible : 'However varied the useful kinds of
labour, or productive activities, it is a *physiological* fact, that they
are functions of the human organism, and that each such function,
whatever may be its nature or its form, is essentially the expenditure
of human brain, nerves, muscles and sense-organs.'[14]

In this sense the physiological similarity of human labour is a
necessary precondition of any value-relation. But only a precondition ! It would be mistaken to regard the physiological concept
of labour as the essence of Marx's theory of value, as many of his

[12] *ibid.* p.165 (72).
[13] *ibid.* 149 (57).
[14] *ibid.* p.164 (71).

critics do. If such an interpretation were correct there would in fact be no fundamental difference between Marx's and Ricardo's versions of the labour theory of value, and the theory itself would be open to serious criticism. In the first place we would have to regard value as a supra-historical category, valid for all economic systems, since in all economic systems labour, looked at physiologically, is only 'an expenditure of human brain, nerves, muscles, and sense-organs.' The essentially historical character of the basic economic categories, on which Marx laid so much stress, would be obscured. And in the second place, we would be compelled, or at least tempted, to look for a mechanical measure of physiological labour, which would, of course, be a fruitless undertaking. (Moreover, we would then really confuse the 'social' with the 'technical', as Joan Robinson thinks.) In fact, however, we have only seen the first part of Marx's solution to the problem so far, since, although labour can be reduced, physiologically, to the simple expenditure of labour-power in any society, such a reduction is only necessary in practice and actually takes place at a particular stage of historical development. This only occurs in a society of commodity owners where exchange constitutes the sole economic connection between individuals, and consequently where commodities are to be regarded as products of average, undifferentiated labour – 'without regard to the form of its expenditure'.[15]

However, this requires no mechanical measure of the physiological expenditure of labour-power, since it is society itself, the spontaneous social process 'behind the backs of the producers', which equates the various forms of labour on the market and reduces them to average 'socially necessary' labour.[16] On the other hand, the 'equality of human labour' in such a society obtains a 'material form . . . in the equal objectivity of the products of labour as values',[17] and only in such a society does 'a definite social relation between men . . . assume the fantastic form of a relation between things'.

What conclusions can be drawn from this short resumé of Marx's theory of value? Clearly that it is impossible to recognise the exclusively social significance of value, unless it is regarded as an historical phenomenon; and that it is equally impossible to deny the historical character of value without seeing in it a 'quality analogous to weight or colour', i.e. a 'technical' category.

This provides the explanation as to why Marx's theory has been

---

15 *ibid.* 128 (38).
16 Cf. p.515ff above.
17 *Capital* I, p.164 (72).

so often misinterpreted by his critics. Incapable of grasping the essentially historical character of economic categories, they simply deduce : If there is such a thing as 'value', then it must be a physical or natural quality of commodities. In this way it is not Marx they criticise, but their own narrowly naturalistic conception!

But what about Joan Robinson's second assertion – her picture of Marx as a seeker for an 'invariable measure of value'? Once more this shows a crassly naturalistic interpretation of Marx's theory.

The first interpreters of the capitalist system, the Mercantilists, asked the question : 'How can the wealth of a nation be reliably measured?' Simply by gold and silver? But the value of gold and silver is itself subject to variation, and a variable measure cannot be an exact measure. So (to give an historical example), the ancient Germans took as a measure of land the area which an average peasant could plough in one day. This was clearly a very imperfect measure; and since then this deficiency has been overcome by the modern technique of surveying. Why couldn't political economy accomplish something similar? It is hardly surprising that William Petty went in search, unsuccessfully, of a 'natural par between Land and Labour, so as might express the value "of all commodities" by either of them alone as well as or better than by both, and reduce one into the other as easily as we reduce pence into pounds'.[18] And Adam Smith expressed the same idea when he wrote : 'Gold and silver vary in their value, are sometimes cheaper and sometimes dearer, sometimes of easier and sometimes of more difficult purchase . . . But as a measure of quantity, such as the natural foot, fathom, or handful, which is continually varying in its own quantity, can never be an accurate measure of the quantity of other things; so a commodity which is itself continually varying in its own value can never be an accurate measure of the value of other commodities.'[19]

But can such an extraordinary commodity of invariable value be found? Smith was convinced that he had found such a charm. In his view the quite ordinary commodity 'labour' could successfully be employed as an 'invariable standard'. Of course the wages paid to workers are usually very different; however, 'equal quantities of labour, at all times and places, may be said to be of equal value to the labourer. In his ordinary state of health, strength and spirits . . . he must always lay down the same portion of his ease, his liberty and his happiness. The price which he pays must always be the same,

---

[18] *The Economic Writings of Sir William Petty* Vol.I, 1899, pp.44-45.

[19] Smith, *An Inquiry into the Nature and Causes of the Wealth of Nations*, 1937 edn., pp.32-33.

whatever may be the quantities of goods which he receives in return of it. Of these, indeed, it may sometimes purchase a greater and sometimes a smaller quantity; but it is their value which varies, not that of the labour which purchases them ... Labour alone, therefore, never varying in its own value, is alone the ultimate and real standard by which the value of all commodities can at all times and places be estimated and compared. It is their real price; money is their nominal price only.'[20]

So much, then, on the historical origins of the concept of the 'invariable measure of value'. It is clear that this insoluble problem (Marx compared it with squaring the circle)[21] only occupied the theorists as long as they regarded 'value' as an eternal natural quality of commodities.[22] The problem is solved, however, as soon as this standpoint is abandoned.

This is not the place to deal with Ricardo's conclusive critique of Smith's fallacy. However, one point should be stressed : whenever Ricardo spoke about the desirability of the so-called 'invariable measure of value' he did not mean 'labour costs', that is the commodity 'labour' which Smith meant, but rather labour as such, as value-creating activity itself, which is, of course, quite a different matter.[23]

However, what does this have to do with Marx and his theory of value? Can we really find traces in his writings which indicate that he might have looked for an 'invariable measure of value', as was certainly the case with Smith, Malthus or Destutt de Tracy? Let us read what Marx himself wrote on this question : 'In order to measure the *value* of commodities – to establish an *external* measure of value – it is not necessary that the value of the commodity in terms of which the other commodities are measured should be invariable. It must

[20] *ibid.* p.33.

[21] *Theories* I, p.150.

[22] Smith regarded 'the creation of value [as] a direct physiological property of labour, a manifestation of the animal organism in man ... Just as the spider produces its web from its own body, so labouring man produces value – labouring man pure and simple, every man who produces useful objects – because labouring man is by birth a producer of commodities; in the same way human society is founded by nature on the exchange of commodities and a commodity economy is the normal form of human economy.' It was left to Marx to recognise that 'value' represents 'a definite social relationship which develops under definite historical conditions'. (Luxemburg, *The Accumulation of Capital*, p.68.)

[23] Cf. the recent commentary on Ricardo's theory of value in R.L.Meek, *Studies in the Labour Theory of Value*, 1956, pp.87, 99, 106-12.

on the contrary be variable, as I have shown in the first part,[24] because the measure of value is, and must be, a commodity since otherwise it would have no *immanent* measure in common with other commodities. If, for example, the value of money changes, it changes to an equal degree in relation to all other commodities. Their relative values are therefore expressed in it just as correctly as if the value of money had remained unchanged. The problem of finding an 'invariable measure of value' is thereby eliminated.' This quotation comes from an extensive discussion on the problem of the 'invariable measure of value' in Marx's *Theories of Surplus-Value*.[25]

But perhaps Joan Robinson didn't consider the *Theories*? Nevertheless she could have found the same ideas in Marx's *Contribution* and in *Capital*. For example, in the *Contribution* : 'Gold must be in principle a variable value if it is to serve as a measure of value, because only as reification of labour-time can it become the equivalent of other commodities, but as a result of changes in the productivity of concrete labour, the same amount of labour-time is embodied in unequal volumes of the same type of use-values.'[26] And we can read in *Capital* : 'But gold can serve as a measure of value only because it is itself a product of labour, and therefore potentially variable in value.'[27]

These few quotations are sufficient to show Marx's real view on this subject. And if it happens to be the case that occasional remarks in Ricardo's works could be interpreted to fit in with Joan Robinson's view (remarks which in no way belong to the essence of his theory of value), then exactly the opposite applies to Marx. Not only did he not share Smith's illusions about an 'invariable measure of value'; he even devoted numerous pages in his *Theories* to an emphatic refutation of this misunderstanding. However, in contrast to so many of his critics, Marx took care not to treat his predecessors as simpletons or small children, but rather showed that even mistakes were necessary steps on the path to the discovery of scientific truth, and that hidden behind the conception of an 'invariable measure of value' lay a very serious and well-founded desire to make the concept of value an objective one. Joan Robinson could have learnt a lot from these pages; she would have found that she already had a predecessor 140 years ago, in the shape of Bailey, who similarly con-

[24] Marx means here his *Contribution to the Critique of Political Economy*, published in 1859.
[25] *Theories* III, pp.133-34. Cf. *Grundrisse*, pp.792ff and, in the German edition, pp.899ff.
[26] *Contribution*, p.67.
[27] *Capital* I, p.192 (98).

fused the idea of an 'invariable measure of value' with the concept of value as an objective social quality of commodities, with labour as its 'inherent' standard of measurement. She would certainly not have then herself characterised Marx as a 'commodity fetishist' . . .[28] But didn't Marx once complain to Engels about his critics : 'If only these people would at least take the trouble to read what I wrote properly !'

2. *Marx's 'rigmarole'*

Up till now we have been concerned with the labour theory of value as it is presented in Volume I of *Capital*. However, Joan Robinson – in common with many much earlier critics of Marx – asserts that there is an 'irreconcilable contradiction' between the labour theory of value in Volume I and the theory of 'prices of production' in Volume III. As soon as she discusses the 'contradiction' she abandons her even temper. 'What was all the fuss about?'[29] – she asks contemptuously in relation to previous debates on this subject. How could Hilferding, Sweezy and other marxists overlook the fact that Marx's attempt at a 'reconciliation' of the two theories 'is purely formalistic and consists in juggling to and fro with averages and totals', and that everything which Marx wrote on this 'is a rigmarole, entirely devoid of content'?[30]

These are serious words. But what real content do they have? To put it plainly : very little. Let us examine Joan Robinson's own words. 'At first Marx states dogmatically that commodities tend to exchange at prices which correspond to their values.'[31] However : 'In a system in which prices correspond to values the net product of equal quantities of labour is sold for equal quantities of money. Thus (given uniform money-wage rates) surplus, in terms of money, per unit of

---

[28] Robinson could reply that she did not in fact mean labour costs, but rather labour-time to be the measure of value. However, this would simply make matters worse, as such an interpretation would signify that value could be directly expressed in units of labour-time, without their having to be compared with one single commodity, which serves as the universal measure of value. We would thus be transported back to the old utopia of 'labour-money', which Marx criticised so remorselessly in the *Grundrisse*.

[29] *Collected Economic Papers*, pp.147-48.

[30] *ibid.*

[31] This is not a correct assertion, as Marx states on p.329 (220) of Volume I of *Capital*: 'We have in fact assumed that prices = values. We shall, however, see in Volume III, that even in the case of average prices the assumption cannot be made in this very simple manner.'

labour is everywhere equal. To say that relative prices correspond to relative values is the same thing as to say that the rate of exploitation is the same in all industries. But if capital per man employed (the organic composition of capital) is different in different industries, while profit per man (the rate of exploitation) is the same, profit per unit of capital must vary inversely with capital per man. It would be possible for both the rate of profit and the rate of exploitation to be equal in all industries only if the ratio of capital to labour employed were also equal.'

She continues : 'In Volume I, Marx leaves this question open. In Volume III he shows that capital per man varies with technical conditions, while competition between capitalists tends to establish a uniform rate of profit. The rate of exploitation therefore cannot be uniform, and relative prices do not correspond to values.' On the contrary, 'the prices of the commodities differ from their values in such a way as to make the rates of exploitation actually enjoyed by the capitalists in the different industries vary with the organic composition of their capitals.' Marx thus 'entangled himself in an artificial difficulty by starting from the assumption of a uniform rate of exploitation. There is no warrant for this assumption. If wages are equal in all industries, surplus per man employed (the rate of exploitation) varies with the net productivity per man employed, and in general, productivity per man is greater where capital per man is greater. In Marx's own words : "The prevailing degree of productive power shows itself in the relative preponderance of the constant over the variable capital . . .If the capital in a certain sphere of production is of a higher composition [than the average] then it expresses a development of the productive power above the average."[32] Thus the rate of exploitation tends to vary with capital per man employed . . . the very same process which produces an equal rate of profit between industries produces unequal rates of exploitation.' 'As I see it', Robinson concludes, 'the conflict between Volume I and Volume II is a conflict between mysticism and common sense. In Volume III common sense triumphs but must still pay lip-service to mysticism in its verbal formulations.'[33]

The entire argument can be dealt with as follows :

1. Marx never maintained that 'to say that relative prices correspond to relative values means the same as to say that the rate of exploitation is the same in all industries'. Neither can such a conclusion be drawn from his theory. This is for one simple reason : Joan

[32] *Capital* III, p.759.
[33] *Essay*, pp.15-16.

Robinson was of course right to say that according to the labour theory of value equal amounts of (average socially necessary) labour are exchanged for equal amounts of labour – eight hours' labour for eight hours' labour, one working day for one working day. But it does not follow from this that the division of the working day into 'necessary' and 'surplus' labour must be the same everywhere. In one case the worker might work, perhaps, five hours 'for himself' and only three hours for the employer; in another case this might be the reverse. But haven't we smuggled in the assumption of uniform wages? This would not improve the matter. Wages may be equal in each of the factories, but the length of the working day – with equal hourly wages – or the intensity of labour can be different. Equal amounts of labour would be exchanged in both cases, and if the organic composition of capital in both plants is the same as the average organic composition of the total social capital, then these values will correspond to prices. But the rates of exploitation could well be different – in contrast to Joan Robinson's assertion. In other words, labour-time as a measure of value is in no way dependent on equal rates of exploitation – and it is amazing, to put it mildly, to see Marx's (or Ricardo's) theory of value interpreted in such a way.

2. On the other hand, Marx never said that because competition leads to the formation of a general rate of profit, 'the rates of exploitation (in different industries) cannot be uniform'. And he similarly never confused the rate of exploitation (i.e. the rate of surplus-value) with 'profits per worker . . . which the capitalists actually enjoy', (i.e. after the originally different rates of profit in different industries have been equalised into a uniform rate of profit). What he actually maintained is the exact opposite : Because equal numbers of workers, who are employed in industries with differing organic compositions of capital, with other conditions being the same (same working time, intensity of labour etc.), produce the same amounts of surplus-value, a transformation of values into 'production prices' must take place, if an average rate of profit is to prevail. The difference is obvious.

3. Marx never maintained that the rate of exploitation would vary with the amount of capital per worker employed – in other words, that the size of the surplus-value produced is a function of the constant capital employed! Such an assertion would have been utterly nonsensical to him. The 'relative predominance of the constant part of capital over the variable' certainly means a growth in the productivity of labour. More commodities, more use-values, can be produced. But that in no way signifies that somehow the workers in industries which employ a larger amount of constant capital automatically create more surplus-value. (An increase in the rate of

surplus-value through increased productivity can only be achieved if the industries producing the means of subsistence are able to supply cheaper consumer goods for the workers, and if the 'necessary' part of the working day is thereby reduced. But this affects all workers in all industries.) Indeed, it would require a lively imagination to ascribe such a 'productivity' theory of surplus-value to Karl Marx![34] We see, then, that it was not Marx himself, but Joan Robinson who 'entangled him in an artificial difficulty', by imputing three axioms to him, which correspond to nothing in his theory. And it is she too who provides him with an easy solution to the 'difficulty' by introducing the only too well-known value-creating power of capital. Just imagine : Marx was perfectly conscious of this solution, yet he still expended years of time and effort to construct the complicated 'rigmarole' in Volume III . . . What a stubborn Hegelian metaphysician he must have been!

3. Marx's search for a social elixir. The problem of value in a socialist society

So much, then, on Marx's theory of value. Finally, however, we are offered some consolation : although, according to Joan Robinson, 'common sense' compelled Marx to admit that the law of values does not function correctly 'under capitalism', he did, how-

---

[34] Robinson cites with approval (from Engels's Preface to Volume III of *Capital*) the view of the Swiss Professor J.Wolf that, according to Marx, 'the production of relative surplus-value rests on the increase of constant capital *vis-à-vis* variable capital', since 'a plus in constant capital presupposes a plus in the productive power of the labourers'. As this view corresponds to Robinson's own it is worthwhile quoting Engels's reply : 'Whenever there is a chance of making a fool of himself over some difficult matter, Herr Professor Wolf, of Zurich, never fails to do so.' And after he has quoted what Wolf has to say, he continues : 'True, Marx says the very opposite in a hundred places in the first book; true, the assertion that, according to Marx, when variable capital shrinks, relative surplus-value increases in proportion to the increase in constant capital, is so astounding that it puts to shame all parliamentary declamation; true, Herr Julius Wolf demonstrates in his every line that he does not in the least understand, be it relatively or absolutely, the concepts of relative and absolute surplus-value.' (*Capital* III, p.15.) As we can see, Robinson had sufficient warning against repeating Wolf's mistake. Despite this, she does not only take over his interpretation, but even chastises Engels for simply 'abusing Wolf without entering into any argument' although 'it is impossible to see wherein Wolf's statement differs from the above statement by Marx' – as if Engels was obliged to deal, in detail, with every crude misunderstanding of Marx's theory.

ever, believe that at least under socialism 'the labour theory of value would come into its own'.[35] In other words, he was clearly a utopian socialist, for whom the labour theory of value did not so much represent the result of pure scientific analysis as rather an artifice to bring about an 'ideal system of pricing'[36] and in so doing secure the realisation of a just world! No wonder that Robinson devotes a particular chapter of her book to Marx's alleged views on the 'Problem of Value in Socialist Society',[37] and primarily to his supposed postulate 'that in a rational economic system prices should be made to correspond to values' : and further that she honestly thinks she has discovered the 'substantial meaning of Marx's theory' in this

[35] *Essay*, p.23.

[36] *ibid.* p.24.

[37] Here is an example of how thoughtlessly Robinson uses passages from Marx. She quotes the following passage from Volume III of *Capital*: 'Only when production is under the conscious and prearranged control of society, will society establish a direct relation between the quantity of social labour-time employed in the production of definite articles and the quantity of the demand for them . . . The exchange, or sale, of commodities at their value is the rational way, the natural law of their equilibrium.' (*Essay*, p.23.) The reader naturally assumes both these sentences relate to the socialist society. But this would be wrong, as Marx, in fact, says exactly the opposite. We read on pp.187-88 of Volume III: 'Every individual article, or every definite quantity of a commodity may, indeed, contain no more than the social labour required for its production, and from this point of view the market value of this entire commodity represents only necessary labour, but if this commodity has been produced in excess of the existing social needs, then so much of the social labour-time is squandered and the mass of the commodity comes to represent a much smaller quantity of social labour-time in the market than is actually incorporated in it. (It is only where production is under the actual, predetermining control of society that the latter establishes a relation between the volume of social labour-time applied in producing definite articles, and the volume of the social want to be satisfied by these articles.) For this reason, these commodities must be sold below their market value, and a portion of them may even be altogether unsaleable. The reverse applies if the quantity of social labour employed in the production of a certain kind of commodity is too small to meet the social demand for that commodity. But if the quantity of social labour expended in the production of a certain article corresponds to the social demand for that article . . . then the article is sold at its market value. The exchange or sale of commodities at their value is the rational state of affairs, i.e. the natural law of their equilibrium. It is this law that explains the deviations, and not vice versa, the deviations that explain the law.'

As we can see, the entire section is concerned with capitalist society, with the exception of the sentence in brackets, in which Marx expresses the view that the future socialist society will not squander the labour-time of its members, as capitalist society does . . . But this does not prevent Robinson from attributing Marx with the view that the sale of commodities at their values will also be the 'natural law' under socialism.

fantasy[38] (shades of Proudhon!). One could perhaps take all this seriously if it at least bore some resemblance to the theory under discussion. But Marx never tired of attacking both Proudhon and all the other utopians who wanted to turn the world upside down by means of a specially constructed 'just system of exchange'. He repeatedly and emphatically stated that value is an historical category, a particular mode of expression of the social function of labour in a society of commodity owners, and that consequently it must, of necessity, disappear in a socialist society.[39]

Thus, in the *Critique of the Gotha Programme* he wrote: 'Within the co-operative society based on common ownership of the means of production, the producers do not exchange their products: just as little does the labour employed on the products appear here *as the value* of these products, as a material quality possessed by them, since now, in contrast to capitalist society, individual labour no longer exists in an indirect fashion but directly as a component part of the total labour.'[40] And in *Capital*: 'The product of labour is an object of utility in all states of society; but it is only a historically specific epoch of development which presents the labour expended in the production of a useful article as an "objective" property of that article, i.e. as its value.'[41] In fact: 'Something which is only valid for this the particular form of production, the production of commodities, namely the fact that the specific social character of private labours carries on independently . . . assumes in the product the form of the existence of value, appears . . . to be just as ultimately valid as the fact that the scientific dissection of the air into its component parts left the atmosphere itself unaltered in its physical configuration.'[42]

This explains why bourgeois political economy 'has never once asked the question why labour is expressed in value and why the measurement of labour by its duration is expressed in magnitude of the value of the product: These formulas, which bear the unmistakeable stamp of belonging to a social formation in which the process of production has mastery over man, instead of the opposite, appear to the political economists' bourgeois consciousness to be as much a self-evident and nature-imposed necessity as productive labour itself.'[43]

---

[38] *Essay*, p.24.
[39] See Chapter 28, pp.428-36 above.
[40] *Critique of the Gotha Programme* in *Selected Works*, pp.319-20.
[41] *Capital* I, pp.153-54 (61).
[42] *ibid.* p.167 (74).
[43] *ibid.* pp.174-75 (80-81).

It is hardly surprising, then, that already during his own lifetime a number of bourgeois academics tried to attribute Marx with the very same view that we have encountered in Joan Robinson's critique, and that he in turn felt obliged to explain that in the course of his investigation of value he 'was concerned only with bourgeois relations, not with the application of the theory of value to a "social state" which Herr Schäffle has constructed for me.'[44]

However, the reader can be certain that the late Austrian Professor Schäffle would not have been a match for Joan Robinson, for in the final analysis he only succeeded in constructing a 'social state' for Marx, whereas Joan Robinson not only created an 'ideal pricing system', but also held out the prospect of 'private saving in a socialist economy' – even, in fact, a socialist income tax and profit tax! But what is one supposed to do with a Keynesian critic who, with unbelievable naïveté turns Marx into a traditional Proudhonist and does not even realise that for Marx 'value' (like nearly all his economic categories) represents not a natural but an exclusively historical category, and that, consequently, Marx never attempted to put together some kind of recipe for the 'socialist kitchen of the future'?

II. MARX'S THEORY OF THE ESSENCE OF CAPITALIST EXPLOITATION AND HIS CONCEPT OF CAPITAL

Up till now we have simply been concerned with Robinson's attacks on the basis of Marx's theoretical system – his theory of value. However, one must be consistent; if we abandon the concept of value, we can no longer retain the concept of surplus-value. With this, we not only destroy the basis, but also the cornerstone of this system – and all the categories of Marx's economics must either be abandoned, or totally revised. This even applies to the apparently simple concept of the 'rate of exploitation', as the ratio s :v is clearly conceived as a value relation. So what is left of Marx's entire system?

What is left, in fact, is the general idea of 'exploitation' and surplus labour – as distinct from surplus-value. This is scarcely enough for the likes of us, but Joan Robinson manages brilliantly with it. She asserts that Marx's 'primitive labour theory of value' has proved to be a complete failure. Nevertheless, 'he used it to express

---

[44] Marx's last economic work, the *Marginal Notes on Adolf Wagner*, *MEW* Vol. 19, pp.360-61.

certain ideas about the nature of the capitalist system, and the importance of these ideas in no way depends upon the particular terminology in which he chose to set them forth'.

And what do these 'ideas' actually consist of? Simply 'that the
possibility of exploitation depends on the existence of a margin between total net output and the subsistence minimum of the workers :
If a worker can produce no more in a day than he is obliged to eat
in a day he is no potential object of exploitation. This idea is simple
and can be expressed in simple language without any apparatus of
specialised terminology.' And 'it is precisely these simple and fundamental characteristics of capitalism' which were explained by Marx
but which became 'lost in the maze of academic economic analysis'.[45]

We see, then, that the 'simple and fundamental characteristic
of capitalism' consists in the existence of surplus labour ! But surplus
labour is as old as the history of human civilisation : 'Capital', says
Marx, 'did not invent surplus labour. Wherever a part of society
possesses the monopoly of the means of production, the worker, free
or unfree, must add to the labour-time necessary for his own maintenance an extra quantity of labour-time in order to produce the
means of subsistence for the owner of the means of production,
whether this proprietor be the Athenian aristocrat, an Etruscan
theocrat, a *civis Romanus,* a Norman baron, an American slave-
owner, a Wallachian boyar, a modern landlord or a capitalist.'[46]

However, it ought to be obvious that if this is all we know about
capitalism, then we know virtually nothing at all, since it is precisely
the 'specific economic form in which unpaid surplus labour is pumped
out of the direct producers', which 'determines the relationship of
rulers and ruled', and also distinguishes the various historical epochs
from one another.[47]

We read in Engels : 'Surplus labour, labour beyond the time
required for the labourer's own maintenance, and appropriation by
others of the product of this surplus labour, the exploitation of labour,
is therefore common to all forms of society that have existed hitherto,
insofar as these have moved in class antagonisms. But it is only when
the product of this surplus labour assumes the form of surplus-value,
when the owner of the means of production finds the free labourer –
free from social fetters and free from possessions of his own – as an
object of exploitation, and exploits him for the purpose of the produc-

---

[45] *Essay,* p.17.
[46] *Capital* I, pp.344-45 (235).
[47] *Capital* III, p.791.

tion of *commodities* – it is only then, according to Marx, that the means of production assume the specific character of capital.'[48]

It is clear, therefore, that Marx's analytical apparatus is the only means by which the specifically capitalist mode of exploitation can be understood; that is, employing the categories of 'value' and 'surplus-value'. It is hardly surprising, then, that Joan Robinson confined herself to the general (and hence totally vague) notion of exploitation as such,[49] without trying to analyse the particular features of the specifically capitalist mode of exploitation. This reminds one of Dühring who 'annexed the surplus labour discovered by Marx, in order to use it to kill off the surplus-value, likewise discovered by Marx, which for the moment did not suit his purpose.'[50] And consequently her conclusions are not much better than those of Dühring.[51]

The best example of this is shown by her treatment of the category of capital. We saw how she accused Marx of not applying his own theory consistently, and how she even imputed a 'fetishistic' concept of value to him. But what did she learn from this epoch-making theory herself? Unfortunately very little. This is because, like Dühring, (and like all present-day 'academic' economists), she regards capital as a thing, as a mere means of production, and not as a social relation. In her eyes it is a natural, not a socio-historical category. So it is not surprising that she rebuked Marx for his 'logic-chopping theorising' in the following way : 'Next Marx uses his analytical apparatus to emphasise the view that only labour is productive. In itself, this is nothing but a verbal point. Land and capital produce no value, for value is the product of labour-time. But fertile land and efficient machines[52] enhance the productivity of labour in

---

[48] *Anti-Dühring*, p.248.

[49] Robinson remarks, with a certain pride, that 'the modern theory of imperfect competition, though formally quite different from Marx's theory of exploitation, has a close affinity with it'. (*Essay*, p.4.) However, in our humble opinion this 'affinity' is about as close as that between the *Communist Manifesto* and the Encyclical *Rerum Novarum*; i.e. it mainly consists in the simple word 'exploitation' which is used both by Marx and the 'modern economists'. The specific nature of capitalist exploitation remains an impenetrable riddle for 'modern theory'.

[50] Engels, *Anti-Dühring*, p.248.

[51] The concept of 'economic surplus' has a different meaning from that used by Robinson. This concept is employed by the American theorists of underconsumption, Baran, Sweezy and Gillman, in place of Marx's concept of surplus-value. We offer no opinion as to whether this is merely a 'change in terminology' (as Sweezy says in a footnote on p.10 of *Monopoly Capital*).

[52] 'Capital' is here suddenly transformed into 'efficient machinery'; as if 'machinery' and 'capital' were synonyms!

terms of real output . . . Whether we choose to say that capital is productive, or that capital is necessary to make labour productive, is not a matter of much importance. What is important is to say that owning capital is not a productive activity. The academic economists, by treating capital as productive, used to insinuate the suggestion that capitalists deserve well by society and are fully justified in drawing income from their property. In the past a certain superficial plausibility could be given to this point of view by treating property and enterprise as indistinguishable. But this method of confusing the issue is no longer effective. Nowadays the divorce between ownership and enterprise is becoming more and more complete . . . The typical entrepreneur is no longer the bold and tireless businessman of Marshall, or the sly and rapacious Moneybags of Marx,[53] but a mass of inert shareholders, indistinguishable from *rentiers*, who employ salaried managers to run their concerns. Nowadays, therefore, it seems simple to say that owning property is not productive, without entering into any logic-chopping disputes as to whether land and capital are productive, and without erecting a special analytical apparatus in order to make the point. Indeed, a language which compels us to say that capital (as opposed to ownership of capital) is not productive rather obscures the issue. It is more cogent to say that capital, and the application of science to industry, are immensely productive, and that the institutions of private property, developing into monopoly, are deleterious because they prevent us from having as much capital, and the kind of capital, that we need.'[54]

We see here yet again that as soon as Joan Robinson begins to criticise she unfailingly misses the mark. It follows automatically from the standpoint of Marx's theory that only labour creates value. But that in no way means that in his eyes, the 'objective factors of production' are to be denied any form of 'productivity'. On the contrary : to the extent that these factors 'raise the level of production' they certainly contribute to the production of use-values (however, that is no reason to confuse the categories of use-value and value as Joan Robinson does). On the other hand, Marx continually stressed[55] that 'capital' (not land) is 'productive' in a different sense : as the ruling social relation of the bourgeois mode of production. We read in the *Grundrisse* and the *Theories* that the 'great historical mission of capital' consists in 'enforcing surplus labour' . . . 'This is why capital

[53] It is pure legend that Marx regarded the capitalist of his time merely as 'the sly and rapacious Moneybags'.
[54] *Essay*, pp.17-19.
[55] Cf. p.220ff above.

is *productive; i.e. an essential relation for the development of the social productive forces.'*[56]

Of course it does not follow from this that capital adds anything to the value of commodities, and that, in this respect, there is no distinction between the activities of the 'factor capital' and the 'factor labour', as Joan Robinson seems to suppose. On the contrary, the difference is enormous; it is no less great than the difference between the activity of a horse, and the 'activity' of the whip which makes it gallop. However, although capital produced no values, it has produced a particular form of exploitation which is indispensable for the development of the productive forces of a particular historical period. And it was in the position to do this precisely because it was 'owned', and not because it somehow functioned as a means of production, or because it furthered the 'application of science to industry'. Its real 'productivity' lies in its insatiable hunger for surplus-value. From this perspective the apparently self-explanatory concept of 'productive labour' acquires a special meaning, since in capitalist society the only labour which is 'productive' is that 'which directly enlarges capital'[57] (or, as Malthus formulated it, 'directly increases the wealth of its master'). Of course 'for a Vulgar Economist this is all a matter of definition' (here I quote Rosa Luxemburg). What is the difference whether we derive the meaning of the word 'productivity' from the relations between person and person, or from the relations between people and nature? The Vulgar Economists never once suspect that the question, 'What is productivity?' has to be looked at historically, and that such a perspective presupposes the use of the dialectical method, to which they object so much.[58]

However, what about the distinction between 'capital' and the 'possession of capital' which Joan Robinson places such weight on? Once more we encounter another old acquaintance, since exactly the same distinction has been the favourite idea of Bray, Gray, Proudhon and other utopian socialists from the year dot.

---

[56] *Grundrisse*, p.325.
[57] *Grundrisse*, pp.305-06.
[58] Luxemburg, *Ausgewählte Reden und Schriften* II, pp.202ff. Moreover: Gillman, the American 'underconsumptionist', in order to establish his theory of 'excess social surplus' in modern capitalism, regards it as necessary to attribute Marx with the view – based on a misunderstanding of a passage from Volume I of *Theories* (pp.396-97) – that 'only such labour is productive whose product is capable of re-entering the cycle of production . . . Thus workers who are engaged in the production of armaments are unproductive in this sense, even though their labour produces products and surplus-value.' It is quite clear that this view has nothing in common with Marx's own.

'If the workers are to be free, then capitalism must be destroyed. But this does not mean the destruction of "capital", but rather its preservation', wrote Bakunin.[59] Marx could only punish such a 'dichotomy' with contempt : We read in the *Rough Draft* : 'Capital . . . is necessarily the *capitalist*. Of course, socialists say, we need capital, but not the capitalist. Then capital appears as a pure thing, not as a *relation of production*.'[60] And in the *Theories* he wrote that when economists speak of the 'services' which capital performs in the production of use-values, they mean nothing other 'than that products of previous useful work serve anew as means of production, as objects of labour, instruments of labour and means of subsistence for the workers . . . But in this sense the word "capital" is quite superfluous and meaningless. Wheat is nourishing not because it is capital but because it is wheat. The use-value of wool derives from the fact that it is wool, not capital. In the same way, the action of steam-powered machinery has nothing in common with its existence as capital. It would do the same work if it were not capital, and if it belonged, not to the factory owner, but to the workers.'[61]

An understanding of Marx's particular concept of capital is, of course, a necessary condition of any discussion of his economic theory.

III. CONCLUDING REMARKS

We have only dealt with the main points of Joan Robinson's critique in this chapter, although she does in fact criticise other parts of Marx's system : his theory of wages, the theory of the falling rate of profit and his theory of crises. Since she does not offer anything new to the marxist reader, however, and what she does say has already been presented with greater effect by other critics of Marx, there is no reason for further discussion on these subjects.[62] And there is equally little reason to correct every misquoted passage and misunderstanding in her presentation of Marx.[63]

[59] Quoted from K.J.Kenafick, *Bakunin and Marx*, 1949, p.92.
[60] *Grundrisse*, p.303.
[61] *Theories* III, p.264.
[62] The main reason why we dealt with Joan Robinson's critique of Marx's law of the falling rate of profit in the 'Appendix' to Part V of this book is because of the influence it has had on the Anglo-Saxon school of marxism (Sweezy, Gillman).
[63] A few examples will suffice:
I. On p.20 of the *Essay* she informs us that, according to Marx, the labour employed in 'packing and preparing commodities for market creates no

On the other hand, a considerable part of her essay is concerned with a discussion of the supposed 'affinity' between Marx's and Keynes's theory (which we consider to be largely imaginary or at least overestimated). Since this does not come under the scope of this chapter we shall confine ourselves to a few concluding remarks.

We have shown how little there is to be learnt from Joan Robinson's critique of Marx! But is this all there is to be found in her book? Doesn't she constantly stress that as a rule the workers are exploited by their employers in present-day society? And doesn't she even attack the capitalist's sacred 'rights of property'? She certainly does do this. She even regards her position as one of the special gains of the 'modern tendency' of political economy – although the 'modern tendency' has to be considerably narrowed down to permit such an interpretation. But this doesn't matter. Joan Robinson should not, at least as an individual, be lumped together with the apologist political economists (including Lord Keynes), but should rather count as a representative of a socialist current in present-day bourgeois economics.

It is of course true that her socialism has its own unique hue. It is heavily reliant on crutches borrowed from the pre-marxist stock of ideas, especially from the godfather of all petit-bourgeois socialisms, Proudhon. And this is not a coincidence as Joan Robinson's socialist conclusions reflect the mood of relatively broad strata of the rebellious bourgeois intelligentsia of today. These strata have lost their belief in the progressive role of the capitalist class; they are deeply disturbed by the 'anti-social practices of the monopolies' and

value'. Exactly the opposite is the case! We read on p.634 of the *Grundrisse*: 'Insofar as trade brings a product to market . . . it gives the product a new use-value (and this holds right down to and including the retail grocer who weighs, measures, wraps the product and gives it a form for consumption) and this new use-value costs labour-time, is therefore at the same time exchange-value.' And Marx says exactly the same in Chapter XVII of Volume III and Chapter VI, Section III of Volume II of *Capital* which Robinson refers to in this context.

II. On p.17 of the *Essay* we read: 'According to Marx's own argument, the labour theory of value fails to provide a theory of prices'. Of course Marx never said this. On the contrary: he referred his readers to the specific 'analysis of competition' which he intended to write, and where the 'real movement of prices would be observed'. (*Capital* III, p.831.)

III. Finally she astounds her readers on p.91 with the discovery that, according to Marx, 'a rise in money wages causes a rise in real wages, and a rise in real wages causes unemployment'. It would be superfluous to quote Marx in reply to this, as everything which he wrote on this subject contradicts this statement.

the economic instability 'of this bedevilled age',[64] and consequently they set their hopes on a nationalised, state-capitalist economy, which would curb the threatening economist chaos, bring about a 'fairer distribution of wealth among the factors of production'[65] and bless us with 'so much capital, and the kind of capital we need'. Hence the sudden dissemination of populist 'Keynesianism' as an ideology which reflects this mood in its iridescent diversity. However, this populist Keynesianism has very little to do with the particular doctrines of Keynes and his school, and one should not hold them responsible for it. Nevertheless – as soon as the academic Keynesians leave their own domain and venture onto the so-called ideological terrain, this unique sub-current of Keynes's economics becomes clearly visible, and we are once more haunted by the ghost of Proudhon! But seen now in this light, the 'socialist' tendencies in Joan Robinson's writing, which so disturbed the late Professor Schumpeter,[66] no longer present anything special or inexplicable.

[64] *Essay*, pp.3ff.
[65] Robinson, *Economics of Imperfect Competition*, p.320.
[66] 'Still more curious [than P.M.Sweezy's book] and a kind of psychological riddle is Joan Robinson's *Essay on Marxian Economics*.' (J.Schumpeter, *History of Economic Analysis*, p.885.)

34.
Neo-Marxist Economics

The wide-ranging, but sadly unfinished text-book by Oskar Lange[1] is to our knowledge the only work in more recent academic marxist literature which consciously, and in detail, takes up the question of the methodology of Marx's *Capital*. This is why the concluding chapter of this book is devoted to this work.

We confine ourselves to the discussion of two questions. That of the object, and that of the method of political economy.

I. A SEEMINGLY DOGMATIC CONTROVERSY

The traditional practice of marxist theory before and after the First World War was to confine the subject matter of political economy merely to the study of the laws of motion of the capitalist, or commodity, economy. This view is rejected both by current Soviet theory and by Western academic theory. In this sense Lange writes : 'Confusion of the concept of the spontaneity[2] of the operation of economic laws with the concept of the objectivity of economic laws has led some economists to the false conclusion that there are no objective economic laws in the socialist formation and that the fact that their spontaneity has been overcome is the result of the fact that they have ceased to operate.' Hence the assertion by these economists 'that in socialist society the science of political economy loses its subject matter. At the most it could only engage in the retrospective examination of pre-socialist formations. Rosa Luxemburg held this view.' Their mistake, concludes Lange, 'is a double one : In the first place they confuse the spontaneity and objectivity of economic laws. And they conclude from the fact that their spontaneity is overcome, that such laws no longer exist at all. This is also the reason why they,

[1] Oskar Lange, *Political Economy* Vol.I, London 1963.
[2] Engels interprets the term 'spontaneous' (*Naturwüchsig*) as 'arising gradually, without intention'.

incorrectly, confine the subject matter of political economy to those relations in which the law of value is operative. And in the second place – in opposition to Luxemburg and Bukharin – the law of value is still valid in the socialist mode of production, although the operation of this law is no longer natural, but rather corresponds to the aims of an organised society.'[3]

These somewhat casually offered criticisms demand a detailed reply.

It is certainly the case that Luxemburg and Bukharin restricted political economy to the investigation of the laws of commodity production. (Except that Lange forgets to add that the same opinion was shared by Hilferding,[4] Schmidt and Boudin, among others.) However, on what basis does Lange impute to Luxemburg and Bukharin the view that there would be no 'objective economic laws' under socialism, and that such laws cannot be found in pre-capitalist societies? He is certainly unable to offer any passage from either author which would allow such a strange interpretation.[5] His only authority is Karl Kautsky, from whose work *Die Materialistische Geschichtsauffassung* he approvingly cites the following long passage : 'This is perhaps a suitable point to draw attention to a mistake which is not uncommon even in socialist circles. It is asserted that it is a peculiarity of commodity production that it operates according to laws. This is supposedly due to the fact that commodity production is carried on anarchically by a great number of producers each of whom disposes of his own means of production. The situation is held to be quite different when society itself takes over the ownership of the means of production. Production can

---

[3] Lange, *op. cit.* pp.84-85.

[4] See his essay, '*Zur Problemstellung der theoretischen Ökonomie bei Karl Marx*', *Die Neue Zeit*, 1904, pp.105, 107.

[5] It is sufficient here to refer to two passages in the *Accumulation of Capital* where Rosa Luxemburg expressly speaks of 'economic laws' which in her view are valid for all human societies. Thus on p.258 she characterises the fact that in the course of history 'living labour is increasingly able to convert more means of production into objects of use in an ever shorter time' as a 'law . . . valid for all economically progressive societies, independent of their historical forms'; and on p.321 we read: 'The formula c greater than v, translated from the language of capitalism into that of the social labour process, means only that the higher the productivity of human labour, the shorter the time needed to change a given quantity of means of production into finished products. This is a universal law of human labour. It has been valid in all pre-capitalist forms of production and will also be valid in the future in a socialist order of society.' These quotations speak for themselves. We assure the reader that the same also applies to Bukharin.

then be organised exactly as society sees fit, quite independently of all economic laws.'

'This,' continues Kautsky, 'is a mistake. If a manufacturer organises a factory, he cannot behave arbitrarily just as he pleases even though he can freely dispose of his own means of production. If certain natural laws of production are not taken into account his enterprise will never be capable of producing anything . . . The difference between capitalist and socialist production is of another kind. In capitalist production it is impossible for the adjustment of production to economic laws to take place without the occurrence of crises. In the socialist mode of production, however, it is possible consciously to adjust production to the natural laws of the mode of production and in this way to maintain the flow of the productive process without catastrophes and crises. This, of course, presupposes that these natural laws are studied. A socialist society which believes that these laws can be opposed by force if only it controls the means of production will aways come to grief.'[6]

We can ignore for the moment the way in which Kautsky reprimands the Bolsheviks.[7] However, we cannot overlook the strange 'natural laws of production' whose existence he asserts with such conviction. It is, of course, in fact true that neither Luxemburg nor Bukharin attached much importance to such laws, for the simple reason that they shared Kautsky's pre-war opinion, according to which the investigation of the 'natural laws of production' is in fact the job of mechanics and chemistry, not political economy.[8]

But what did Luxemburg and Bukharin actually think? Why did they cling to the idea that political economy's sole concern is the investigation of the laws of commodity production? One thing is certain, it was not for the reason which Lange ascribed to them. It is sufficient to read through a few pages from Rosa Luxemburg's *Einführung in die Nationalökomonie* to appreciate this: She asks,

[6] Kautsky, *Die Materialistische Geschichtsauffassung* Vol.I, 1927, pp.876-77.

[7] This passage, which was omitted from the Polish translation used by Lange, reads: 'The Bolsheviks, who thought it was enough to become masters of the means of production in order to run the economy as desired, have paid dearly for their mistake – or rather it was the Russian people which had to pay the penalty. As the old saying runs: When the kings (or dictators) rave, the people get the beating.'

[8] Kautsky wrote, 'Marx's intention in *Capital* was to investigate the capitalist mode of production . . . He did not concern himself with natural laws, which are the basis of production; their study is the task of mechanics and chemistry, not political economy.' (Karl Kautsky, *Karl Marx' Ökonomische Lehren*, 2nd Edition, 1906, p.3.)

can there be a 'universal' science of political economy which can be applied, with equal validity, to capitalism, as well as to pre-capitalist societies? She answers no, because in contrast to the production relations of capitalism, those of pre-capitalist societies were so 'self-evidently simple and transparent' that they did not require 'dissection by the scalpel of political economy'. What is immediately obvious in the study of such societies is that 'there need determines and channels labour so directly, and the result corresponds so precisely to the intention and the need', that 'all connections, causes and effects, labour and its result are entirely open to view . . . One can twist and turn the economy as one likes, one finds no riddles which can only be discovered by means of a deep analysis, by means of a special science.' Of course this economy can and must constitute the object of research into its sociology and economic history;[9] but a special economic theory does not seem to be appropriate here.

The situation is quite different with the economics of capitalism, as we read further on in Rosa Luxemburg's book : 'If we look into one individual private enterprise, a modern factory, or a huge complex of factories and works, such as Krupp, or at an agricultural ranch in North America, then we find there the strictest organisation, the most extensive division of labour, the most refined forms of planning based on scientific findings. Everything works marvellously – directed by one will, one mind. However, no sooner have we left the gates of the factory or the farm, than we are greeted by chaos. Whereas the numerous individual parts are highly organised, the whole of the so-called "people's economy" (*Volkswirtschaft*), i.e. the capitalist world economy, is totally disorganised. In this totality, which envelops oceans and continents, no plan, consciousness, or regulation makes itself felt; only the blind rule of unknown, unfettered forces plays its capricious game with the economic destiny of mankind . . . And it is precisely this', concludes Rosa Luxemburg, 'which produces the unpredictable and puzzling result which makes

---

[9] Luxemburg wrote : 'The most stupid peasant knew in the Middle Ages that his dire situation had a simple and direct cause; first, the unlimited extraction of taxes and forced labour by the landlord; second, the theft by the same lords of common land, woods, meadows and rivers. And the peasant pronounced what he knew to all the world in the peasant wars . . . The only matter here which remained to be scientifically explained was the question of the historical source and development of those relations, the question as to how it could come about that throughout Europe the formerly free rural estates were transformed into medieval lordships, to which interest and taxes were due, and the formerly free peasants into a mass of subjects, first obliged to provide forced labour and later bound to the land.' (*Einführung in die Nationalökonomie*, in *Ausgewählte Reden und Schriften* Vol.I, p.470.)

the economy into an alien phenomenon, estranged to us, and independent of us, whose laws we must discover in the same way as we investigate the phenomena of nature, as we seek to discover the laws which govern the life of the plant and animal kingdom, the changes in the crust of the earth and the movements of the atmosphere.'[10] This, then, is Rosa Luxemburg's view. With all the best will in the world it is impossible to discover in her statements the 'confusion of spontaneity and objectivity' which Kautsky and Lange impute to her : even less so, as the pages from her *Einführung* which we have cited, are essentially a paraphrase of the train of thought which can already be found in Marx's *Capital*. Marx teaches that what characterises bourgeois society is that there is no *a priori*, 'conscious social regulation of production'. It is therefore a society in which the social relations of production confront people as alien, reified and ruling powers, and in which 'what is rational and necessary . . . can only assert itself as a blindly working average'.[11] And the form in which it asserts itself is that of 'automatically operating' 'natural laws of society' of production and exchange,[12] independent of the will of people, which in the first instance remain unknown to the producers

---

[10] Luxemburg, *ibid.* pp.464, 468-69, 480-81.

[11] Letter to Kugelmann, 11 July 1868, *Selected Correspondence*, pp.195-97.

[12] The way in which Lange interprets the marxist concept of 'natural laws of society' is interesting. In his opinion all that Marx wishes to express by the term 'natural law' is that this is a question of 'iron laws', 'independent of human will' – i.e. objective economic laws. (Lange, *Political Economy*, pp.57-58, note 18.) And because all economic laws – be they in a capitalist, pre-capitalist or socialist society – have this character of objectivity, then the economic laws of all societies can and must be regarded as 'natural laws'. (This creates a bridge to 'eternal' supra-historical economics.) However, Marx in fact characterised as 'natural laws' only those economic connections which impose themselves 'as a blind law upon the agents of production' instead of being 'understood and controlled by their common mind' (*Capital* III, p.257) – that is, only the laws of commodity production, and above all of capitalist production, for only the latter form of production exhibits economic conditions 'which assert themselves without entering the consciousness of the participants and can themselves be abstracted from daily practice only through laborious theoretical investigation'. (*ibid.* p.899.) George Lukacs demonstrated in *History and Class Consciousness* (pp.231-32) that this is the real meaning of Marx's 'natural laws of society'.

The same interpretation of natural laws as that of Lange can be found in the Soviet philosopher, Rosenthal, in whose book *Die Dialektik in Marx' 'Kapital'* we read: 'Marx stresses with the use of the concept of "natural-historical processes" the fact that processes in society as well as in nature are conditioned by objective laws.' (pp.43-44.) This is yet another instance where we can see the tendency towards absolutising Marx's essentially dialectical concepts.

themselves and must be discovered and deciphered *post festum*. Of course this is true only as long as social development resembles a 'process of natural history' and society consequently requires a special science, whose task it is to advance in the manner of the natural sciences, from the phenomena on the surface of economic life to the 'inner law' of these phenomena, to their 'hidden inner essence'.[13] Thus, it is only the reified and mystified form of the bourgeois relations of production, their apparent natural law-like behaviour which, in Marx's view, requires scientific explanation and which constitutes the *raison d'être* of the specific science of political economy.

However, stresses Marx, 'The whole mystery of commodities, all the magic and necromancy that surrounds the products of labour on the basis of commodity production, vanishes . . . as soon as we come to other forms of production.' Marx means here primarily the 'Asiatic and ancient etc. modes of production' in which 'the transformation of the product into a commodity, and therefore men's existence as producers of commodities, plays a subordinate role', and for this reason they appear 'as extraordinarily simpler and more transparent' than the mode of production of capital.[14] However, the same simplicity also characterises the feudal society of the middle ages for the reason that in this form of society 'personal dependence forms the given social foundation, there is no need for labour and its products to assume a fantastic form different from their reality . . . The corvée can be measured by time just as well as the labour which produces commodities, but every serf knows that what he expends in the service of his lord is a specific quantity of his own personal labour-power. The tithe owed to the priest is more clearly apparent than his blessing. Whatever we may think, then, of the different roles in which men confront each other in such a society, the social relations between individuals in the performance of their labour appear at all events as their own personal relations, and are not disguised as social relations between things, between the products of labour.'[15]

[13] Classical economy, stressed Lukacs, 'with its system of laws is closer to the natural sciences than to any other. The economic system whose essence and laws it investigates does in fact show marked similarities with the objective structure of that Nature which is the object of study of physics and other natural sciences. It is concerned with relations that are completely unconnected with man's humanity . . . Man appears in it only as an abstract number, as something which can be reduced to number or to numerical relations. Its concern, as Engels put it, is with laws that are only understood, not controlled.' (*op. cit.* p.232.)

[14] *Capital* I, p.172 (76).

[15] *ibid.* p.170 (77).

And finally the same amazing transparency is also offered by the 'association of free individuals' of the future, 'expending their many different forms of labour-power in full self-awareness as one single labour force' : 'The total product of the association is a *social* product. One part of this product serves as fresh means of production and remains social. But another part is consumed by the members of the association as means of subsistence. This part must therefore be divided amongst them. The way this division is made will vary with the particular kind of social organisation of production, and the corresponding level of social development attained by the producers.' Although this may vary, the social form does not offer anything of a secretive nature : 'The social relations of the individual producers both towards their labour and the products of their labour are here transparent in the simplicity in production as well as in distribution.'[16]

The critics of Luxemburg might well agree that Marx did indeed contrast the 'simplicity' and 'intelligibility' of the relations of production of all non-capitalist societies with the 'mystical veil'[17] which shrouds capitalist relations of production; and it is also true that one can find many passages in Marx which view the specific task of political economy as being the investigation of the capitalist economic order.[18] But does it follow from this that we can manage without a theory of political economy of non-capitalist societies, as Rosa Luxemberg supposed? Engels apparently held a different opinion ! He wrote, in *Anti-Dühring* : 'Political economy, in the widest sense, is the science of the laws governing the production and exchange of the material means of subsistence in human society. Production and exchange are two different functions. Production may occur without exchange, but exchange – being necessarily an exchange of products – cannot occur without production.' And in conclusion : 'The mode of production and exchange in a definite historical society, and the historical conditions which have given birth to this society, determine the mode of distribution of its products.' And further : 'The conditions under which men produce and exchange vary from country to country, and within each country from generation to generation. Political economy, therefore, cannot be the same for all

---

[16] *ibid.* pp.171-72 (79).
[17] *ibid.* p.173 (80).
[18] The very title of Marx's work, *Critique of Political Economy*, refers to the fact that Marx did not regard his task as being the refutation of this or that school or opinion in political economy, but of the whole of previous political economy as the theoretical reflection of the capitalist mode of production.

countries and for all historical epochs . . . Anyone who attempted to bring the political economy of Tierra del Fuego under the same laws as are operative in present-day England would obviously produce nothing but the most banal commonplaces. Political economy is therefore essentially a historical science. It deals with material which is historical, that is, constantly changing; it must first investigate the special laws of each individual stage in the evolution of production and exchange, and only when it has completed this investigation will it be able to establish the few quite general laws which hold good for production and exchange in general.'[19]

At first sight this seems to contradict Rosa Luxemburg's view; but to what extent? In order to answer this question we first have to agree upon the meaning of what Engels wrote. Engels defined political economy as a science of the laws 'which govern production and exchange', but at the same time pointed out that there could be societies without exchange (e.g. 'primitive communism' or the future socialist society). Engels's definition therefore seems *prima facie* simply to say that the object of political economy cannot be extended beyond societies with exchange (i.e. commodity-producing society). And this is the reason why Lange finds it necessary to 'correct' Engels by simply declaring that what Engels really meant is not 'exchange', but the 'distribution' of products among the members of society, and that we consequently have to define political economy as a science 'of the laws of production and distribution'![20] (However, Lange does not notice that such an interpretation would only precipitate us into new difficulties; for, as according to Engels, distribution is determined by the relations of production and exchange, we would be driven to the awful conclusion that distribution is determined by distribution!)

Nevertheless, let us leave such hair-splitting casuistry on one side! Those who do not approve of Engels's definition, those who feel that it is too narrow, are certainly free to replace it with another – to the effect that political economy 'in its widest sense' not only has the task of studying the economic relations of societies with exchange, but also societies without exchange, i.e. all human societies. But it is doubtful what would be gained by such a reinterpretation of Engels's defini- for the proponents of 'supra-historical economics'. This is because the very same Engels says directly on this point that – as an 'essen-

---

[19] *Anti-Dühring*, pp.177-78.

[20] 'Frederick Engels defined political economy as the science "of the laws governing the production and exchange of the material means of subsistence in human society". This agrees entirely with our definition. We have only replaced the term "exchange" by the term "distribution".' (Lange, *Political Economy*, p.6, note 6.)

tially historical science' – political economy primarily has to concern itself with the investigation 'of the special laws of each individual stage of development of production and exchange', and then only right at the end 'establish the few general laws which hold good for production and exchange in general'. The terrain of 'universal' (supra-historical) political economy is thereby brought down to a minimum and its significance substantially reduced. So it is not surprising when Lange complains that 'Engels does not fully appreciate the significance of this branch of political economy'.[21]

It must be admitted that Lange's recourse to Engels has not proved to be particularly convincing. But do we really have to devalue the classic works of socialism by treating them like the Holy Scriptures? Marx and Engels were only human, and therefore had the privilege of erring! Instead of relying on this or that 'text' we ought rather to learn from the experience of present-day Soviet economics, which has made several attempts to create a handbook of political economy in its 'widest sense'. Can these attempts be regarded as successful? Hardly. What they offer the reader is merely an amalgam of incoherent parts – of the economic history of pre-capitalist social formations, the economic theory of capitalism, as provided by Marx, and the descriptive-normative theory of the present-day Soviet economy. All this can certainly be characterised as the 'science of economics' in its broadest sense. And it is certain that neither Rosa Luxemburg nor Bukharin would have disputed it. They merely asserted that we do not require a particular economic theory of socialism and pre-capitalist social formations – on the lines of the theories of Ricardo and Marx. So, finally, the whole controversy seems to dissolve into a purely terminological dispute.

However, this terminological difference is in fact an appearance which conceals a very real difference. Lange and other economists of the 'Eastern bloc' know only too well that the social and economic order, whose interpreters they are, can in no way claim to have overcome the reification and law-like nature of economic phenomena, and that in the interests of its self-preservation this economic structure must, in fact, do its utmost to create the greatest possible space for market forces within the overall context of central planning. Consequently what these economists strive for is a narrow and specialised discipline of 'state-economics', a 'socialist public-finance' which – following the example of the economic theory of the West – accepts the categories of commodity, money and the market as things which

[21] *ibid.* p.95, note 2.

are the eternally given data of economic life[22] and which consciously rejects the 'utopian' idea of the 'simplicity' and 'intelligibility' to be sought in socialist relations of production. And if the advocates of this view still appeal to Marx and Engels this is only intended to accommodate the letter of marxism to a social practice which becomes and must become increasingly distant from its spirit.

II. ON THE METHOD OF MARX'S ECONOMICS

1. It is evident that if one does not wish to confine the task of economic theory to the study of capitalist society alone, but rather, instead of this, strives for the creation of an economic theory of all succeeding social formations – and if one simultaneously hankers after a 'timeless', 'universal' political economy – then one would also select a methodology which corresponded to this end and which can be equally applied to the relations of production of monopoly capitalism, and to those of Tierra del Fuego. However, in this case the specific methodology of Marx's *Capital* would necessarily come off second best and would have to be replaced by more or less useful academic discussions on the method of economic science 'as such'.

Oskar Lange devotes no less than three chapters of his work to questions of methodology. We have already looked at one of these chapters (on the nature of 'economic laws') in the preceding part of our critique, and it is unnecessary to come back to it now. The second contains a (somewhat dubious) presentation of the materialist conception of history; however, since sociology cannot serve as a substitute for political economy, the methodological value of this chapter is very questionable. What remains is the third chapter which deals directly with the 'Method of Political Economy'. Unfortunately this chapter also offers the reader no more than the popular presentations

---

[22] This is naïvely put by the Polish economist Temkin, according to whom Marx's 'polemical' economic theory has to be transformed into a 'positive' and 'constructive . . . theory of the socialist economy'. 'It became clear in the 1930s that even in a developed socialist society commodity and money relations cannot be completely superseded. One therefore adjusted oneself to this by the fact that although central planning and the market represented opposing economic forms, they should complement and correct one another.' The issue is then to discover 'how, with the retention of central planning as the power which determines general social and economic goals, market forces can fulfil the role of the economic stimulus and the determinant of details of economic development'. (Temkin, *Karl Marx' Bild der kommunistischen Wirtschaft*, Warsaw 1962, pp.24-25.)

of marxist economics which existed previously . . . In fact, we discover from it that Marx – in opposition to the majority of bourgeois economists – proceeded not only from social man (instead of from man 'as such'), but also from the social man of a particular historical period, and it is precisely this which separates his economics from present-day 'academic' economic science. These findings are, however, not especially novel, and above all, they fail to reveal the methodological assumptions which enabled Marx to bring about this epoch-making revolution in the science of economics.[23] In other words : What is missing from Lange's methodological chapter is the 'soul' of Marx's method of political economy – his dialectic !

2. But isn't this a mere manner of speaking intended to annoy troublesome opponents – a piece of ritual, incomprehensible even to those who pretend to understand it?

At least for Marx himself the question of the application of the dialectic to the area of economic theory was of decisive importance ! This can be seen from the numerous critical comments on Ricardo's methodology, which can be found in Marx's works. The question primarily revolves around the role of abstraction in political economy. 'Ricardo', states Marx, 'consciously *abstracts* from the form of competition, from the appearance of competition, in order to comprehend the *laws as such*'. However, 'he must be reproached for not going far enough, for not carrying his abstraction to completion; . . . on the other hand one must reproach him for regarding the phenomenal as *immediate and direct proof* or exposition of the general laws, and for failing to interpret it. In regard to the first, his abstraction is too incomplete; in regard to the second, it is formal abstraction which in itself is wrong . . . The vulgar mob has therefore concluded that theoretical truths are abstractions which are at variance with reality, instead of seeing, on the contrary, that Ricardo does not carry true abstract thinking far enough and is therefore driven into false abstraction.'[24]

How should we interpret these critical remarks on Ricardo's method? Why are the abstractions which he uses to be regarded as on the one hand 'not far-reaching enough', and on the other as merely 'formal', i.e. as artificial? As far as the first criticism is concerned, we can find numerous examples of this. Let us recall the shortcomings of Ricardo's theory of value. In the first place, the theory is almost

---

[23] 'The results', wrote Engels, 'are nothing without the development which led to them – we already know that from Hegel.'

[24] *Theories* II, pp.106, 437.

exclusively concerned with the relative size of the value of commodities, but not the substance of value, i.e. value itself. Correspondingly, what is missing in Ricardo is any investigation of the specific character of value-creating labour – as distinct from those qualities which are ascribed to labour 'as the creator of use-values'.[25] On the other hand there is also no recognition of the fact that value-creating labour (although this is private labour in each concrete instance), must represent itself as its opposite, as universal social labour (which naturally presupposes the exchange of the products of labour, i.e. a historically specific mode of production).[26] Therefore Ricardo also fails to understand that 'exchange-value', which for him is the crucial issue, is simply a mode of appearance of value, and that the development of the value-relation itself must push towards this form, and finally, to the formation of money.[27]

These deficiencies in Ricardo's theory of value certainly bear witness to a 'deficient power of abstraction', the incapacity of the classical economists to see the qualitative aspect of the problem of value behind its quantitative aspect, and the substance of value behind the form of its appearance. In fact all these deficiencies can be reduced to a common denominator which consists in the fact that Ricardo (like all the classical economists) overlooked the most essential thing – the specific social form of value-creating labour, and naïvely equated this labour with human labour in general.[28]

Thus, according to Marx, it was the class-based limitation of the economics of Smith and Ricardo which in the final analysis resulted in its own 'lack of the theoretical understanding needed to distinguish the different forms of economic relations'.[29] Or, expressed

[25] *Capital* I, p.132 (41) and p.312, note 2 (p.204, note 1). That this is in no way mere hair-splitting is proved by the fact that it was only possible for Marx to discover the crucially important categories of constant and variable capital, organic composition of capital etc. on the basis of his differentiation of the 'twofold character of labour'.

[26] 'Ricardo's mistake is that he is concerned only with the magnitude of value. Consequently his attention is concentrated on the relative quantities of labour which the different commodities represent, or which the commodities as values embody. But the labour embodied in them must be represented as social labour, as alienated individual labour . . . This transformation of the labour of private individuals contained in the commodities into uniform social labour, consequently into labour which can be expressed in all use-values and can be exchanged for them, this qualitative aspect of the matter . . . is overlooked by Ricardo.' (*Theories* III, p.131.)

[27] Cf. p.124 above.
[28] Cf. *Capital* I, pp.173-74 (80-81).
[29] *Theories* I, p.92.

in terms of method, because the specific bourgeois forms of production appeared as immutable natural forms to the Classical economists, and because they proceeded from them as given presuppositions, it was in their interests not to interpret them 'genetically', but rather to 'seek to reduce them by means of analysis to their inner unity', i.e. to the law of value.[30] Consequently they had to regard the economic forms of the bourgeois mode of production as 'something merely formal which did not affect its content', the production of use-values, of goods.[31] Thus the methodological problem of the conflict between 'content' and 'form' could not arise for the classical economists. And this is the point where the dialectic comes into its own. For, according to the dialectical conception, each 'content' and the 'form' which it gives rise to are in constant interaction and struggle with one another – which on the one hand results in a shedding of the forms, and on the other a transformation of the content.[32] By contrast, if 'form' is regarded as something accessory, external to the content, then it is inevitable that either the form will be neglected, sacrificed to content (as in the case of the classical economists), or the attempt will be made to turn this form into an absolute. We can take as an example of the latter approach those Soviet economists who conclude from the fact that even a socialist society will have to distribute the amounts of social labour at its disposal, and measure it by labour-time, that the economic category of value will also prevail under socialism, i.e. from the supra-historical substratum of the determination of value they infer the supra-historical character of the form of value. It is clear then that the methodological importance of the dialectic for marxist economics cannot be estimated too highly![33]

The undialectical elements in the theoretical analyses of Ricardo and the classical economists can be seen, on the other hand, in their 'methodical avoidance of the categories of mediation',[34] in their efforts to deduce the phenomena on the surface of economic life 'by simple abstraction directly from the general law or to show by cunning argument that they are in accordance with that law'.[35] We know that,

---

[30] *Theories* III, p.500.

[31] *ibid.* p.54.

[32] Lenin said that one of the fundamental elements of the dialectic is the 'struggle of content and form, and conversely. The throwing off of the form and the transformation of the content.' (*Collected Works* Vol.38, p.222.)

[33] See Chapter 3 above on the significance of the problem of content and form for Marx's methodology.

[34] Lukacs, *op. cit.* p.176.

[35] *Theories* I, p.89.

according to Marx 'all science would be superfluous if the outward appearance and the essence of things directly coincided'.[36] However, in reality, 'the final pattern of economic relations as seen on the surface, in their real existence and consequently in the conceptions by which the bearers and agents of these relations seek to understand them, is very much different from, and indeed quite the reverse of, their inner but concealed essential pattern and the conception corresponding to it'.[37] Lukacs writes, commenting on this sentence by Marx : 'If the facts are to be understood, this distinction between their real existence and their inner core must be grasped clearly and precisely. . . . Thus we must detach the phenomena from the form in which they are immediately given and discover the intervening links which connect them to their core, their essence. In so doing we shall arrive at an understanding of their apparent form and see it as the form in which the inner core necessarily appears.'[38]

This explains the fundamental importance of 'transitions' and 'intermediary links' (i.e. the above-mentioned 'categories of mediation') for Marx's methodology ! Without these categories (which resemble the so-called process of approximation of academic theory, only in outward appearance, but which in fact represent a materialist 'inversion' of Hegel's dialectical method), Marx's *Capital* would have been inconceivable. It is therefore clear that Marx had to criticise Ricardo on this point (and precisely on it), and reproach him for the 'formal' and 'arbitrary' manner of his abstraction.

In fact Ricardo already unexpectedly introduced the hypothesis of the general rate of profit in the first chapter of his work, dealing with 'value' – in order to show that even this assumption does not contradict the determination of the value of commodities by labour-time, and merely constitutes an 'exception'. Marx comments on this : 'Instead of postulating this general rate of profit, Ricardo should have rather examined how far its existence is in fact consistent with the determination of value by labour-time, and he would have found that instead of being consistent with it, *prima facie* it contradicts it, and that its existence would therefore have to be explained through

---

[36] *Capital* III, p.817. This passage could have just as well come from Hegel's *Logic*, which, in Volume II continually counterposes the world 'as it appears' and the world in 'its being-in-itself' and sees the 'truth of appearance' in 'essence'. (*Science of Logic* Vol.II, p.142.)

[37] *Capital* III, p.209. ('The distinction between idea and concept is also to be found in Hegel.' Lukacs, *op. cit.* p.25.)

[38] Lukacs, *ibid.* p.8.

a number of intermediary stages, a procedure which is very different from merely including it under the law of value.'[39]

However, it is just this elaboration which is absent in Ricardo! No wonder, then, that the question, 'how from the mere determination of the "value" of the commodities their surplus-value, the profit and even a general rate of profit are derived remains obscure' to him.[40]

'Where he correctly sets forth the laws of surplus-value he distorts them by immediately expressing them as laws of profit. On the other hand, he seeks to present the laws of profit directly, without the intermediate links, as laws of surplus-value',[41] in the same way that his method generally 'omits some essential links and *directly* seeks to prove the congruity of the economic categories with one another'.[42] 'One can see that though Ricardo is accused of being too abstract, one would be justified in accusing him of the opposite : lack of the power of abstraction, inability when dealing with the values of commodities, to forget profits, a factor which confronts him as a result of competition.'[43] And exactly the same can be said about the other parts of his work – on his conception of capital, wage-labour, money etc. The *Grundrisse* states on this : 'He never investigated the form of the mediation.'[44]

However, despite all this, stresses Marx, the method of investigation employed by Ricardo is 'justified', and its 'scientific necessity in the history of economics' cannot be denied![45] For what Ricardo intended with this method, and where for the most part he succeeded, was 'to reduce the various fixed and mutually alien forms of wealth' (profit, interest, rent) 'to their inner unity', that is, to understand 'the inner structure of the bourgeois economic system . . . in contrast to the multiplicity of its outward forms'. Certainly, Ricardo's theory, 'occasionally contradicts itself in this analysis. It often attempts to carry through the reduction directly, leaving out the intermediate links, and to prove that the various forms are derived from one and the same source. This is however a necessary consequence of its analytical method, with which criticism and understanding must

[39] *Theories* II, p.174.
[40] *ibid.* p.190.
[41] *ibid.* p.374.
[42] *ibid.* p.165.
[43] *ibid.* p.191.
[44] *Grundrisse*, p.327.
[45] *Theories* II, p.164.

begin. Classical economy is not interested in elaborating how the various forms come into being, but seeks to reduce them to their unity by means of analysis, because it starts from them as given premises. But analysis is the necessary prerequisite of genetical presentation, and of the understanding of the real, formative, process in its different phases.'[46]

Marx here sketched out the essential difference between his and Ricardo's method of investigation, a difference which is related to the different role allotted to analysis in the theoretical systems of the two thinkers. Of course, both are analysts, because the fundamental identity of the economic categories, their congruence with one another, can only be demonstrated through analysis. However, whereas for Ricardo, who regarded the social forms of bourgeois production as 'given forms', analysis represents the Alpha and Omega of his method, for Marx it is only a necessary step in the process of scientific discovery and must, therefore, be complemented and expanded by 'genetic' investigation,[47] the task of which is to investigate the processes of the development and formation of the economic categories themselves, and their further development through the various phases. But what is in fact analysis enriched by the genetic method of investigation? Nothing other than Marx's dialectical method!

3. Marx's remarks on Ricardo's method are therefore important for us because we believe that they provide a key to the understanding of Marx's *Capital*. A twofold methodological task stood before Marx in the light of his criticisms of Ricardo : on the one hand, to discover through scientific abstraction those categories and concepts by means of which the most essential relations of the capitalist mode of production could be understood, i.e. the 'essence' in contrast to the mere 'forms of appearance'; and on the other hand, to connect these essential relations with the phenomena on the 'surface' of economic life, or rather to derive these phenomena from the essential relations.

Marx accomplished the former task by excluding from his analysis, in the first instance, all the phenomena of competition etc., in order to limit the study exclusively to 'capital in general', i.e. the

---

[46] *Theories* III, p.500.

[47] It was in this sense that Hegel characterised 'objective logic' (the doctrine of Being and Essence) as the 'genetic exposition of concepts' (subjective logic). (*Science of Logic* Vol.II). Cf. Lukacs, *op. cit.* p.175 : 'this ability to go beyond the immediate . . . means the transformation of the objective nature of the objects of action.'

production and circulation process of capital in its pure form;[48] in the course of this he studied, in a genuinely dialectical manner, the individual economic forms, not in a static sense, but in the flow of their movement, and not only from the perspective of the individual capital, but also (and primarily) from the perspective of aggregate social capital. This method was the only one by which he could investigate the 'life history' of capital, i.e. not simply uncover the inner laws which govern its actual operation, but also its development as a category from money and value, and in addition the developmental tendencies which point beyond its form of production. Only after this task had been carried out was it possible to proceed, via numerous 'mediations' and intermediary stages, to 'capital in reality' – the representation of the competition of capitals, the credit system etc. (A whole series of tasks which is commenced in Volume III of *Capital*, but which, according to Marx's plan, was to have been concluded with the theory of the formation of prices, the industrial cycle and crises, presupposing a representation of the world market.)[49]

4. The reader, whose patience has been so sorely tried, may well now be saying : All this may be correct; but what has it got to do with Lange's book? The reproach is justified. Lange's book in fact contains nothing, or almost nothing, on the method of Marx's *Capital*. Why should he bother to deal in detail with the role played by 'abstraction', 'progressive concretisation' and 'verification' in the methodology of political economy as such, if he does not mention the special importance of this method of reasoning in marxist economics?[50]

This criticism should not be misunderstood : If Lange's book had been published in 1909, rather than 1959, the lack of a chapter on methodology could not have occasioned such criticism. At that time the attention of marxist theoreticians was so totally absorbed with the material, with the concrete content of Marx's work that even the most important of them (with the exception of Lenin, Luxemburg and the young Hilferding)[51] scarcely gave any attention to the unique method of Marx's economic work, or at the least 'left it in the back-

---

[48] Cf. Chapter 2, pp.41-50 above.

[49] Cf. Chapter 2 above, where the question of the structure of Marx's *Capital* is dealt with in detail.

[50] With the exception of two quotations from Grossmann and the Polish economist W.Brus.

[51] We refer here to Hilferding's sketch of the history of ideas in *Die Neue Zeit*.

ground'. This was of course no accident, since theoreticians of the Second International, who for the most part were oriented towards Neo-Kantianism and positivism, had lost any feeling for the philosophical point of departure adopted by Marx and Engels, i.e. Hegel's conceptual schemes.[52] Thus Lenin could write in his *Philosophical Notebooks* of 1914-15 : 'It is impossible completely to understand Marx's *Capital*, and especially its first chapter, without having thoroughly studied and understood the *whole* of Hegel's *Logic. Consequently, half a century later none of the marxists understood Marx! !*'[53]

Since that time (i.e. the period before the First World War) more than five decades have elapsed, and something could have been learnt on this subject in the intervening years. For example, in 1923 George Lukacs published *History and Class Consciousness*, which taught us to look at Marx's method with quite different eyes. Ten years later Lenin's *Philosophical Notebooks* were published; their significance for marxist economics cannot be overated. Furthermore, the heyday of Soviet economics in the 1920s provided many valuable methodological discoveries – to name only the works of Preobrazhensky, and

[52] The following section from Otto Bauer's review of Hilferding's *Finanzkapital* is very characteristic in this respect. (*Der Kampf*, 1909-10, p.392.): 'Along with Marx's method he also took over Marx's manner of presentation – even the Anglicisms.' (Hilferding's manner 'of imitating the inimitable language of Marx' was later mocked by Preobrazhensky in his book *Das Papiergeld in der Epoche der proletarischen Diktatur.*) Bauer continues : 'This appropriation of Marx's presentation is not entirely without its dangers. Marx, as always with the foundation of a new science, developed an entire system from graphic images, comparisons, metaphors, tropes and symbols, and in these he clothed his laws and concepts. We often forget that we ourselves speak in images when we say, for example, that the value of the means of production is "transferred" to the commodity produced, that value is "expressed" in price, that the law of value "appears" in the movement of prices . . . The present tendency of the science of our time, however, is to proceed in the manner of presentation from such variegated images to abstract concepts. Fulfilling this need is necessary, not so much because Marx's image-filled language, which arose under the influence of Hegel's similarly metaphorical language, has misled several authors into reinterpreting marxism as a metaphysical system, but rather because this manner of presentation is not the one appropriate to the present day – because it obstructs the victorious progress of the marxist system.'
One can see that for Bauer, Marx's dialectic (e.g. the doctrine of 'essence' and 'appearance') is nothing more than 'metaphorical language' taken from Hegel . . . No wonder that he saw this dialectic – in this in accord with 'the science of our time' – as superfluous ballast.
[53] Lenin, *Collected Works* Vol.38, p.180.

the Rubin[54] school.[55] And finally, the publication of Marx's *Rough Draft* for *Capital* in 1939-40 amounted to a veritable revelation, which introduced us, as it were, to Marx's economic laboratory, and laid bare all the fine details and complex byways of his methodology. Since that time one no longer has to bite into the sour apple and 'thoroughly study the whole of Hegel's *Logic*' in order to understand Marx's *Capital* – one can arrive at the same end, directly, by studying the *Rough Draft*. Without the appropriation and thorough assimilation of the methodological discoveries of *Rough Draft*, it is impossible, in our opinion, to make any real progress in the field of marxist economics. But unfortunately, in the light of this, O. Lange's otherwise useful and interesting book makes a remarkably old-fashioned impression!

III. CONCLUDING REMARK

Every historical period and society has the theory 'which it deserves'. However, marxist social science has more than three decades of unparalleled degradation and sterility behind it; only half freed from the Stalinist strait-jacket, it has to learn the difficult art of free thought and free speech.[56] Is it then any surprise that it is still far from attaining again the relative high point of the 1920s?

[54] We read the following on Rubin's school in a work by the Soviet philosopher Rosenthal: 'Rubin's adherents and the Menshevik idealists, who spread their mischief in the 1920s and 1930s into the fields of political economy and philosophy, have written a great deal on the "dialectic of capital", but they treated Marx's revolutionary method in the spirit of Hegelianism, and turned it into a scholarly game of concepts, a complex system of artifice and intricacy, far remote from science . . . The Communist Party has destroyed this tendency, which is quite alien to marxism, and assisted Soviet philosophers and economists to unmask its essence.' (East German edition of 1957, p.19.) As we know, the Rubin school was in the main 'destroyed' by the execution of Rubin and his comrades in Stalin's concentration camps and prisons . . . Soviet philosophers would be better advised at least to keep silent about this painful subject, rather than make such comments.
[55] The only publication in the West which can be named in this connection is Marcuse's *Reason and Revolution*, 1941.
[56] Lange's treatment of Stalin's alleged contribution to sociology and political economy is typical in this respect. As late as 1959 he still feels obliged to pay tribute to the dead dictator and point out his scientific 'merits', by allotting quite disproportionate space to his views, in particular to his so-called law 'of the necessary conformity between production relations and the character of the productive forces'. Admittedly only in the text. In the footnotes

But not only that! The degradation of marxist theory, which we have been able to observe in the last decades, was of course no accident, no whim of history (as the belated critics of the so-called personality cult wish us to believe). It was a phenomenon which necessarily accompanied the far-reaching structural changes in society which Lange very prudently identifies as the predominance 'of the conservative interests of certain strata or social groups whose position is the result of the place they occupy in the superstructure'.[57] He says one thing, but means another : he talks of the 'superstructure', but means the state and party bureaucracy.[58] Nevertheless, no matter how these strata are defined, the actual pressure which the 'conservative interests' exercise is not thereby made any the less. And this pressure does not only find its expression in the economy (primarily in the relations of distribution), as well as in the omnipotence of the 'greatest fetish of all' – the state – but also in the spheres of science, culture, art, social morality etc. And the more the 'conservative interests' seek to convince themselves and their associates of the stability of the prevailing situation – gloomily sensing the precarious and provisional nature of their historical position – the stronger the urge to seek all kinds of 'eternal values' in life, thought and emotions. Hence the tendency towards absolutising and de-historicising the old marxist heritage in philosophy, ethics, sociology (the theory of the state), economics etc. Marx's materialism does not seem to represent any insuperable obstacle to such an interpretation (although they look askance at the philosophical heritage of the young Marx). However, the situation is different with Marx's dialectic, which 'includes in its positive understanding of what exists a simultaneous recognition of its negation, its inevitable destruction' and 'regards every historically developed form as being in a fluid state'.[59] This dialectic of contradictions is naturally instinctively repugnant to the 'conservative interests'; hence the efforts made to pay lip-service to it, to force it into the

---

he did not neglect to add that Stalin only discovered the name of the law because it had already been 'discovered and formulated by Marx and Engels'. What would one say about a theologian who lauded his God in the text, and denied his existence in the footnotes? !

[57] Lange, *Political Economy*, p.82, note 57.

[58] The concept of bureaucracy as a social stratum has unpleasant 'oppositional' overtones; however, one is permitted to speak of bureaucratic habits, of individual officials seized by bureaucratic ways, but not of the bureaucracy as a social stratum. Accordingly the most recent Soviet textbooks define bureaucracy as a 'remnant of pre-socialist administrative methods'. (*Fundamentals of Marxist Philosophy*, in Russian, 1960, p.535.)

[59] *Capital* I, p.103 (20).

Procrustean bed of eternal natural laws (where it can create less mischief), and at the same time to banish it from the social, economic and political theory (and practice) of the present day. 'Eternal values' are difficult to reconcile with the critical-revolutionary dialectic of Marx.

What is important is to acknowledge this connection and to struggle against the pressure of 'conservative interests' in every sphere. Only in this way will it be possible to go beyond 'neo-marxism' (or more correctly : 'vulgar-marxism') both in sociology and the economics.

Bibliography

Works by Marx and Engels

Marx K. and Engels, F. *Collected Works*, London, Lawrence and Wishart 1975.

—*Communist Manifesto* (1848), in *Selected Works in One Volume*, London, Lawrence and Wishart 1968.

—*The German Ideology* (1845-46), London, Lawrence and Wishart 1965.

—*Perepiska*. (*Marx and Engels's Correspondence with Russian Political Figures*) (in Russian), Moscow 1947.

—*Selected Correspondence*, Moscow, Progress Publishers 1975.

—*Selected Correspondence* ed. Dona Torr, London, Martin Lawrence 1934.

—*Selected Works in One Volume*, London, Lawrence and Wishart 1968.

—*Werke*, Berlin, Dietz Verlag 1965-73.

Marx, K. *Capital*, Vols. I, II and III, London, Lawrence and Wishart 1970, 1974.

—*Capital*, Volume I (transl. Ben Fowkes), Harmondsworth, Penguin Books, 1976.

—*Contribution to the Critique of Political Economy* (1859), London, Lawrence and Wishart 1971.

—*Critique of the Gotha Programme* (1875), in *Selected Works in One Volume*, London, Lawrence and Wishart 1968.

—*Economic and Philosophical Manuscripts of 1844* (transl. Martin Milligan), London, Lawrence and Wishart 1970.

—*Grundrisse der Kritik der politischen Ökonomie (Rohentwurf)* (1857-58), Berlin, Dietz Verlag 1953. Includes Marx's 'Excerpt Notes' on Ricardo and Bastiat and Carey, and *Fragment des Urtextes 'Zur Kritik der politischen Ökonomie'* (1858) a preliminary draft for the *Contribution to the Critique of Political Economy*.

—*Grundrisse, Foundations of the Critique of Political Economy (Rough Draft)*, (transl. Martin Nicolaus), Harmondsworth, Penguin Books 1973.

—*Marginal Notes on Adolph Wagner's 'Lehrbuch der politischen Ökonomie'* (1879-80), in Marx-Engels *Werke (MEW)* Vol. 19, pp.355-383. Transl. by Athar Hussain in *Theoretical Practice*, Issue 5, Spring 1972.
—*The Poverty of Philosophy* (1847), New York, International Publishers 1963.
—*Pre-Capitalist Economic Formations*, ed. and introduced by E. J. Hobsbawm, London, Lawrence and Wishart 1964.
—*Selected Writings*, eds. T. Bottomore and M. Rubel, Harmondsworth, Penguin Books 1963.
—*Theories of Surplus-Value* Parts I, II and III, London, Lawrence and Wishart 1969.
—*Wage-Labour and Capital* (1847), in *Selected Works in One Volume*, London, Lawrence and Wishart 1968.
—*Wages, Prices and Profit* (1865), in *Selected Works in One Volume*, London, Lawrence and Wishart 1968.
Engels, F. *Anti-Dühring*, London, Lawrence and Wishart 1955.
—*Selected Writings*, ed. W.O. Henderson, Harmondsworth, Penguin Books 1967.

Other works cited

Basso, L. 'Rosa Luxemburg: The Dialectical Method' in *International Socialist Journal*, November 1966.
Bauer, O. *Rezension über Marxliteratur* (Review of literature on Marx) in *Der Kampf*, Vol. 6. 1913, p.190.
—*Die Akkumulation des Kapitals* in *Die Neue Zeit*, 1913, No. 23.
—*Kapitalismus und Sozialismus nach dem Weltkrieg*, Vol. I: *Rationalisierung-Fehlrationalisierung*, Vienna, 1931.
—*Die Nationalitätenfrage und die Sozialdemokratie*, 2nd edition. Vienna, 1934.
—Review of Hilferding's *Finanzkapital* in *Der Kampf*, 1909-10.
Behrens, F. *Zur Methode der politischen Ökonomie. Ein Beitrag zur Geschichte der politischen Ökonomie*, Berlin, 1952.
Bernstein, E. Review of Hilferding's *Böhm-Bawerk's Criticism of Marx* in *Dokumente des Sozialismus*, Heft 4, 1904.
—*Zur Theorie des Arbeitswerts* in *Die Neue Zeit*, Nos. 12 and 13, 1899.
Bigelow, J. *Jamaica in 1850: or the effects of Sixteen Years Slavery*, New York, 1851.
Birkenffeld, L. *'Die Konsolidierung der sozialistischen Arbeiterinternationale'*, in *Archiv für die Geschichte des Sozialismus und der Arbeiterbewegung*, 1930.

Block, H. *Die Marxsche Geldtheorie*, Jena, 1926.

Böhm-Bawerk, E. von, *Karl Marx and the Close of his System*, with an Introduction by Paul Sweezy, Clifton NJ, Kelley 1973.

Bortkiewicz, L. von, 'Value and Price in the Marxian System' in *International Economic Papers* No.2, 1952. (Originally published in German, 1906-07).

Bukharin, N. *Imperialism and the Accumulation of Capital*, London, Allen Lane the Penguin Press 1972.

Bulgakov, S. *On the Question of Markets in the Capitalist Mode of Production*, (in Russian), Moscow, 1897.

Burns, Sir Alan, *History of the British West Indies,* 1954.

Diehl, K. *Sozialökonomische Erläuterungen zu David Ricardos Grundgesetzen der Volkswirtschaft und Besteuerung.* Leipzig, 1905.

Grigorovici, T. *Die Wertlehre bei Marx und Lassalle. Beitrag zur Geschichte eines wissenschaftlichen Missverständnisses.* Vienna, 1910.

Grossmann, H. *'Die Änderung des ursprunglichen Aufbauplans des Marxschen "Kapital" und ihre Ursachen'* in *Archiv für die Geschichte des Sozialismus und der Arbeiterbewegung (Grünbergs Archiv)* Vol. 14. 1929.

—*Das Akkumulations– und Zusammenbruchsgesetz des Kapitals,* Frankfurt, 1967.

—*'Marx, die klassische Ökonomie und das Problem der Dynamik'* (mimeographed), New York. English translation 'Marx, Classical Political Economy and the Problem of Dynamics', *Capital and Class* 2 and 3, 1977.

—Review of Sternberg's *Imperialism* in *Archiv für die Geschichte des Sozialismus und der Arbeiterbewegung (Grünbergs Archiv),* 1928.

Hegel, G.F.W. *Science of Logic*, London, Allen and Unwin 1929.

—*Encyclopädie der philosophischen Wissenschaft im Grundrisse,* Leipzig, 1870.

—*'Die Verfassung Deutschlands'*, in *Politische Schriften* ed. Habermas, Frankfurt, 1966. (Quoted in Lukacs, G. *The Young Hegel*).

Hilferding, R. *'Zur Problemstellung der theoretischen Ökonomie bei Marx'*, in *Die Neue Zeit* 1904, No. 4.

—*Das Finanzkapital*, Frankfurt, 1968.

—*Böhm-Bawerk's Criticism of Marx*, Clifton NJ, Kelley 1973.

Kautsky, K. *Karl Marx's ökonomische Lehren*, Stuttgart, 1887.

—*Die materialistische Geschichtsauffassung*, Berlin, 1927.

Kenafick, K. *M. Bakunin and Karl Marx*, 1949.

Kowalik, T. *Rosa Luxemburg's Economic Theory* (in Polish, mimeographed) Warsaw, 1963.

Lange, O. *Political Economy*, Vol. I *General Problems*, New York, 1963.

Lassalle, F. *Die Philosophie Herakleitos des Dunklen von Ephesos*, Berlin, 1858.

Lenin, V.I. *Collected Works*, London, Lawrence and Wishart 1960-70.

—'*Marginal Notes on Luxemburg's "Accumulation of Capital"*' (in Russian), *Lenininskii Sbornik*, Vol. 22.

Leontiev, L. *O pervonatshalnom nabroske 'Kapitala' Marksa*, Moscow, 1946.

Liebknecht, W. '*Erinnerungen an Marx*' in *Ausgewählte Schriften*, Vol. I, 1934.

Lukacs, G. *History and Class Consciousness*, London, Merlin 1971.

—*The Young Hegel*, London, Merlin 1975.

Luxemburg, R. *The Accumulation of Capital*, London, Routledge and Kegan Paul 1963.

—*The Anti-Critique* in *Imperialism and the Accumulation of Capital*, London, Allen Lane 1972.

—*Ausgewählte Reden und Schriften*, 2 Vols. Berlin, 1951.

—*Briefe an Freunde*, Hamburg, 1950.

—'*Einführung in die Nationalökonomie*' in *Ausgewählte Reden und Schriften*, Vol. I, Berlin, 1951.

Malthus, T.R. *Principles of Political Economy*, London, 1836.

Marcuse, H. *Reason and Revolution*, London, Routledge and Kegan Paul 1941.

Mattick, P. *Rebels and Renegades*, Melbourne, 1946.

Mayer, G. *Friedrich Engels*, Berlin, 1920-33.

Meek, R.L. *Studies in the Labour Theory of Value*, London, Lawrence and Wishart 1956.

Morf, O. *Das Verhältnis von Wirtschaftstheorie und Wirtschaftsgeschichte bei Karl Marx*, Berne, 1951.

Moszkovska, N. *Zur Kritik moderner Krisentheorien*, Prague, 1935.

—*Das Marxsche System. Ein Beitrag zu dessen Ausbau*, Berlin, 1929.

—'*Zur Verelendungstheorie*', in *Die Gesellschaft*, 1930.

Oppenheimer, F. *Wert und Kapitalprofit*, Jena, 1916.

Petty, W. *The Economic Writings of Sir William Petty* Vol. I, London, 1899.

Preiser, E. '*Das Wesen des Marxschen Krisentheorie*' in *Wirtschtft und Gesellschaft Festschrift für Franz Oppenheimer zu seinen 60. Geburtstag.* Frankfurt, 1924.

Preobrazhensky, E. *The New Economics*, London, Oxford University Press 1965.

Ryazanov, D. *Karl Marx: Man, Thinker, and Revolutionist*, London, Martin Lawrence 1927.

—'Siebzig Jahre "Zur Kritik der politischen Ökonomie"' in *Archiv für die Geschichte des Sozialismus und der Arbeiterbewegung (Grünbergs Archiv)* Vol. 15, 1930.

Ricardo, D. *Principles of Political Economy and Taxation*, (3rd edn. 1821) Harmondsworth, Penguin Books 1971.

Robinson, J. Introduction to R. Luxemburg, *The Accumulation of Capital*, London, Routledge and Kegan Paul, 1963.

—*An Essay on Marxian Economics*, London, Macmillan 1966.

—'The Labour Theory of Value: A Discussion', in *Science and Society*, 1954.

—*Collected Economic Papers*, Oxford, Blackwell 1960.

—*Economics of Imperfect Competition*, London, Macmillan 1933.

Rodbertus-Jagetzov, C. *Schriften*, Berlin 1898.

Rosenthal, *Die Dialektik in Marx' 'Kapital'*, Berlin, 1957.

Rubin, I.I. *Essays on Marx's Theory of Value*, Detroit, Black and Red 1972.

—*Marx's Theory of Production and Consumption* (in Russian) 1930.

Schlesinger, R. *Marx, His Time and Ours*, London, Routledge and Kegan Paul 1950.

Schmidt, A. *The Concept of Nature in Marx*, London, New Left Books 1971.

Schumpeter, J. *Capitalism, Socialism and Democracy*, London, Unwin 1966.

—*History of Economic Analysis*, New York, 1954.

Sismondi, J.C.L. *Political Economy*, Clifton NJ, Kelley 1970.

Smith, A. *An Inquiry into the Nature and Causes of the Wealth of Nations* (1776), New York, 1937.

Sternberg, F. *Marx und die Gegenwart*, Cologne, 1955.

Strachey, J. *Contemporary Capitalism*, New York, 1956.

Sward, K. *Legend of Henry Ford*, New York, 1948.

Sweezy, P. *Theory of Capitalist Development*, London, Dobson 1946.

Sweezy, P. (and Baran, P.) *Monopoly Capital*, New York, 1964.

Temkin, G. *Karl Marx' Bild der kommunistischen Wirtschaft*, Warsaw, 1962.

Trotsky, L. *The Revolution Betrayed*, New York, Pathfinder Press 1972.

—*Permanent Revolution*, New York, Pathfinder Press 1976.

Tugan-Baranovsky, M. *Studien zur Theorie und Geschichte der Handelskrisen in England*, Jena, 1901.

—*Theoretische Grundlagen des Marxismus*, Leipzig, 1905.

Index